Microgrid Recordings of Dynamic Constant Power Loads using Analytical Stability Method

TUSHAR SINGH

FOREWORD:

Distributed energy resources(DER) that use renewable energy have become more prevalent due to environmental challenges and the depletion of fossil fuel reserves. Increased penetration of DER in a distribution system allows for establishing a local independent system. These local systems connected with DER are known as a microgrid. The advantages of microgrids include increased reliability, lower energy costs, reduced greenhouse gas emissions, and the capacity to help populations during protracted power outages.

DC microgrid is a distribution system and an emerging electrical network on ships, aeroplanes, satellites, and electronic devices. The DC microgrid applications have grown rapidly in the power industry as usage of renewable energy sources (RES), DC intrinsic loads, and energy storage devices have increased.

Renewable energy resources usually operate at a low voltage. Therefore, a power electronic-based converter is required in a DC microgrid to convert the voltage from the renewable source to the voltage acceptable for the DC bus/load. The research shows that these power electronic converters generate constant power loads (CPLs). These CPLs shows nonlinear behaviour and have negative impedance characteristics. Maintaining system stability under varying operating conditions due to unpredictable nonlinear CPLs is difficult for present DC microgrid architecture. Hence, one of the critical challenges with DC microgrids is the nonlinear CPL caused by power electronic converters. Therefore a controller design approach addressing nonlinearity is needed to minimise the adverse effects of CPLs.

Most prior research has been directed toward developing linear or nonlinear control with ideal CPLs that may not be practical. This dissertation presents a systematic approach to develop an intelligent adaptive robust controller that ensures desired voltage control and outlines a simple method for determining the stability conditions for DC microgrids having nonlinear dynamic CPLs.

The primary contribution of this research is developing a nonlinear controller design that enhances DC microgrid voltage control and reduces the adverse effects caused by unpredictable nonlinear CPLs. The suggested control method is achieved by an adaptive robust control (ARC), which successfully integrates adaptive control and deterministic robust control (DRC) design methodologies. The design approach uses the H-infinity Cubature filter (HCKF) based observer for estimating the uncertain nonlinear power and is used in control design. The proposed method enhances performance by maintaining the benefits of both adaptive control and DRC. In particular, the method ensures superior performance for transient errors in the presence of uncertain parameter variations (during fault and sudden load changes) for nonlinear CPLs.

This thesis first presents a robust control architecture for a DC microgrid made up of a photovoltaic (PV) system and an energy storage system(ESS) established on linear matrix inequality (LMI). The controller is designed based on the Hardy space infinity control method to maintain the bus voltage constant. This has been accomplished by creating a reliable controller based on LMI while maximising the stability region.

The design approach considers the nonlinearities contributed by the CPLs and achieves H-infinity performance. Secondly, an ANFIS-based adaptive controller is designed, where uncertain parameters are considered. These parameters are computed using the cubature Kalman filter (CKF) approach, which is appropriate for the nonlinear design. Finally, these strategies are combined to create adaptive robust control, which significantly increases system effectiveness and efficiency.

The suggested controller designed for a DC microgrid with dynamic CPLs has a large operating range, quick dynamic response, and accurate tracking, in addition to ensuring global stability. The closed-loop system's asymptotical stability is assured to the desired equilibrium, and specific conditions are developed to direct the design approach. Both real-time simulation case studies and stability analysis are carried out for each control strategy established in this thesis to validate their efficacy. The simulation results are compared with those obtained using traditional methods to demonstrate their superiority.

At last, this dissertation includes hardware-in-the-loop testing to verify the effectiveness of the study. HiL testing combines rooftop PV and energy storage and has been validated with the WAVECT control platform.

The suggested controller provides better performance compared with the conventional control design in settling time, percentage overshoot, and steady-state error during initial transients, large load changes, and fault conditions. The proposed design method achieves a fast settling time of 0.12s compared to a traditional PI controller, estimated at 1.6091s after a fault. In terms of steady-state error, the approach performs superior with 0.11% error, while for the traditional PI controller, it is estimated as 0.9231%. The overshoot of 0.22 percent is also less compared to the traditional PI controller, which is 0.326 percent.

TABLE OF CONTENTS

CHAPTER 1

INTRODUCTION :

1.1 Introduction and Background:

The future of energy generation and transmission is redefined because of ecological challenges. As a result of environmental challenges, several countries have begun to shift towards a more sustainable energy supply based on renewable sources [1]. Although fossil fuels have long been the primary source of electric power generation, renewable energy-based power generation is gaining popularity as a way to lessen carbon emissions from fossil fuel-powered generating stations. Many countries have been transitioning their energy systems in recent years from a more centralised strategy of continuous fossil fuel-based generation to a more decentralised approach of variable generation from thousands of power plants[2].

In a centralised generation, the energy is generated at large-scale facilities, which are usually located away from end-users. Examples of centralized generation include fossil-fuel-based power plants, nuclear power plants, hydroelectric dams, wind farms, etc. Decentralised or distributed generation, on the other hand, refers to a collection of sources and other devices that generate power near consumption points. Photovoltaic systems, wind energy systems, microturbines, fuel cells, and diesel engines are examples of distributed generation[3],[4].

The requirement for a workable power system to deliver a clean, safe, and constant supply of electricity has become more important with the need for reliable power demand, the gradual depletion of fossil fuels, and the growing challenge of environmental degradation. Hence, the use of distributed energy resources is growing rapidly and will account for around 30% of global gross electricity generation in the near future, enabling the development of a more environment-friendly power system with higher energy efficiency and reduced pollutant emissions[5]-[8].

Distributed energy resources, commonly identified as RESs (for example, solar and wind energy), are challenging to direct connection to the utility grid owing to their intermittent nature. Consequently, the new age energy system has to be flexible and reliable to accommodate renewable energy supplies. A microgrid is a new type of electricity distribution system that connects distributed energy resources to local demands. It has gained appeal as a means to use RESs and boost the reliability of power systems. [9]-[11]. Microgrids can be defined as low-voltage distribution networks made up of a range of small-scale distributed generation units, for example, wind and solar energy systems, energy storage units, and various loads [12]. These microgrids can have various advantages compared to conventional energy distribution systems, including increased dependability, improved power quality, and lower distribution network losses.

These microgrids can be categorised as AC or DC microgrids. Earlier microgrids used AC to replicate a traditional power system, but advances in modern power electronics have given DC a higher chance of survival. Compared to an AC microgrid, a DC counterpart can attain better efficiency, greater dependability, freedom from frequency issues and coordination, freedom from three-phase unbalance, and energy storage can be easily connected[13],[14]. This is possible because DC microgrids offer superior efficiency by eliminating AC-DC and DC-AC conversion phases. In practice, many renewable distributed energy sources are basically DC sources. In addition, new DC loads like computers, plug-in hybrid electric vehicles, data servers, and variable speed drives have made DC systems increasingly desirable in recent years[15].

These advantages increased the DC microgrid's suitability for various applications, including data centre supply, rural electrification, railway and naval power systems, and building power solutions. Also, DC Microgrids are an answer to the issue of grid resilience and reliability to physical disturbances. The majority of these applications make use of DC microgrids, an active research area right now. Researchers face several challenges with DC microgrids in each application mentioned above[16]. A DC microgrid's control and stability are fundamental challenges, among many other challenges that must be overcome to operate effectively. Each microgrid must maintain stable voltage operation by implementing proper control of its distributed energy resources (DERs), especially during fault and sudden load change conditions using its controllable components, to realise the benefits of DC microgrids.

The conditions listed below highlight the difficulties with microgrid stability and control:

- Unpredictable load switching, faults, and their consequent changes to the electric network, along with the randomly variable power generation from distributed renewable energy sources (RESs), produce significant disturbances in the microgrid voltage.
- DERs with power converter interfaces within a microgrid replace synchronous generators, reducing ride-through capability due to their low kinetic energy storage capacity and rotational inertia reduction.

- The high level of interaction between various DG units, each of which has a different interface and controller, can negatively affect the microgrid's stability.

Power engineers must create novel control strategies to maintain the electric power system's stability in the face of these problematic uncertainties [17]. Most DC microgrid loads operate as a constant power load because modern power electronics devices interface with them. Constant power loads (CPL) are loads for which the output power from the converter is constant, i.e., the power output and input are equal when the power dissipated into the circuit is ignored. In other words, a minor rise in the current causes the output voltage to drop. This indicates that the CPL exhibits incrementally negative impedance characteristics because the relative rate of change linking the current and voltage is less than zero ($dv/di < 0$).

A system's stability point indicates that the system will return to that point in the event of a disturbance. However, the system connected to CPL cannot return to this state if there is any disruption. Therefore, CPLs with a negative impedance impact may result in serious stability concerns[18]-[20]. Despite the fact that earlier research suggested several novel strategies for DC microgrid voltage control during disturbance conditions, these strategies are unable to maintain control during large disturbances.

The control design approach needs to use straightforward but efficient adaptive control methodologies to create a genuinely robust microgrid in an unpredictably changing environment. Unfortunately, until now, this has not been entirely feasible, and that is the topic of this dissertation. This dissertation aims to address the inadequacies mentioned above by proposing an ANFIS-based adaptive robust control methodology that can handle the instability brought about by massive signal disturbances. Lyapunov-based stability analysis for the suggested control design is also provided to demonstrate the approach's efficacy.

Following the literature review, motivation, objectives, and outcomes of the dissertation, this introductory chapter presents the concept of DC microgrids and the associated control challenges in brief.

1.2   Control Challenges and Literature Review

DC microgrid has recently attracted much attention to increasing the power supply's efficiency and dependability. The literature on DC microgrids from a control perspective is fast growing due to the increased interest in DC microgrids. A DC microgrid can function in two modes: islanded and grid-connected. A microgrid, in a grid -connected mode, is capable to link to the main grid to improve power delivery. Although a DC microgrid can upgrade the urban electricity distribution in grid-connected mode, its primary application is supplying power in an emergency because of its capacity to function in an islanding manner.

The microgrid is separated from the primary grid and must depend on itself to serve local loads in islanded mode. Studies and real-world examples demonstrate that an islanded DC microgrid can successfully address the problem of power supply in small rural or remote areas, where these communities are unable to connect to the utility grid for technical and economic reasons. Utilizing an islanded DC microgrid power generation system that takes advantage of the available renewable energy resources is a more economical method of electrifying an off-grid neighbourhood.

In islanding mode, the microgrid keeps operating on its own, even if disconnected from the grid. Because the main grid does not support it, the microgrid in islanded mode faces control issues due to multiple nonlinear loads connected to it. Nonlinear loads can cause voltage and current distortions and can be challenging to manage in a DC microgrid that is functioning in islanded mode. This is because the microgrid must balance the power produced by its various sources and the power consumed by its loads while maintaining stable voltage. This can be particularly challenging if there are multiple nonlinear loads connected to the microgrid, as they can interact with each other in unpredictable ways.

There are numerous studies in the literature that discuss various control designs of DC microgrids[21]-[23]. The first section focuses on the control challenges of DC microgrids operating in islanded mode and connected to energy storage and distributed energy source systems. The next section briefs a review of the DC microgrid's control architecture in islanded mode.

1.2.1 Control Challenges for DC Microgrid:

The successful working of a DC microgrid in islanded mode requires a steady voltage and acceptable transient performance under various loading scenarios. Inadequate load supporting and disturbance-damping capability is the primary cause of instability and poor transient performance. Microgrids typically feature a significant number of power electronic devices for cascade, parallel, drive, and isolation, which are together known as a multi-converter power electronic-based system, to fulfill the essential necessities of varied loads[24],[25].

These converters must be subjected to some controls to keep system characteristics (voltage and power sharing among various generators) at desirable levels or to adjust these values during transient or unexpected load increases[26]-[28]. Hence, the converter is crucial for the control scheme of the DC microgrid. The load converter exhibits the characteristics of a CPL when control is applied tightly. These CPLs have negative incremental impedance, contributing to system instability. The negative impedance of output across the source converter's terminals decreases the system damping. Thus, a limit cycle characteristic is produced by the output of source subsystems tending to an internal oscillation between the capacitor and inductor connected to the source converters. This characteristic causes an unstable dc-bus voltage and raises the control stress on source converters [29].

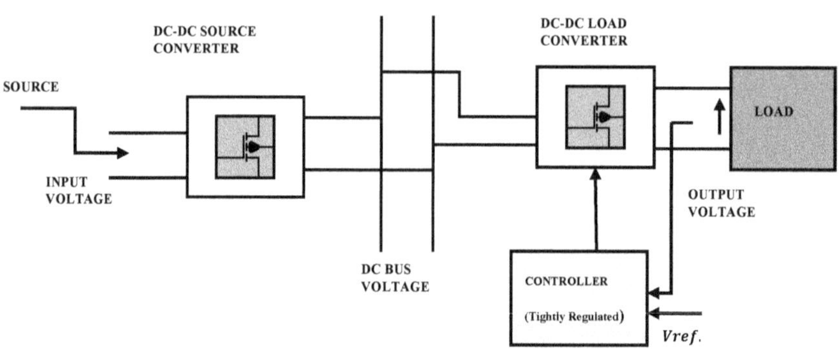

Fig.1.1: Power electronic-based multi-converter system connected to CPL

A CPL adjusts its impedance in response to changes in input voltage to keep a constant output power. A switching regulator is an example of a load that consumes constant power. The regulator must draw the same measure of power from its source to maintain the load power constant even if the source's voltage varies. This presents a negative impedance characteristic since the current must increase if the voltage decreases to maintain the output power constant. Some other loads, such as electronic loads, LED lights, and motor drives with tightly regulated controllers, behave as CPL. Also, most DC-DC converters with resistive loads behave like a CPL.

As displayed in Fig. 1.1, a load is linked to the DC bus using an interface that actively controls its output power. As a result, the load will have a piecewise constant power output independent of voltage or current disturbances. Since nonlinear CPLs show a negative impedance behavior, the voltage of a CPL will decrease as the current injection is increased, unlike constant current or impedance loads. Negative impedance effectively reduces the system's resistive damping, which can lead to more severe stability issues. A huge CPL, for example, could cause a DC microgrid's voltage level to drop excessively.

Furthermore, a poorly damped DC microgrid may readily encounter excessive voltage oscillations when subjected to disturbances.

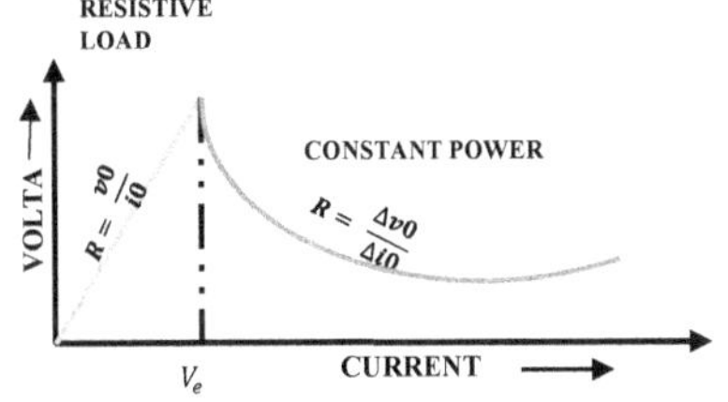

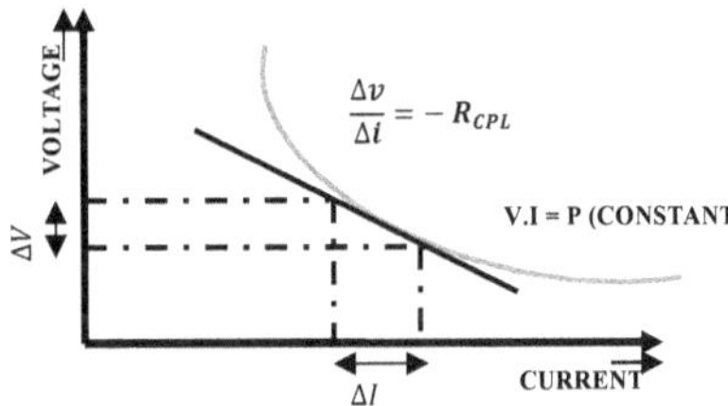

Fig.1.2: V-I characteristics of CPL   **Fig.1.3: Negative impedance characteristic of CPL**

In practice, the converter behaves as a resistive load in an open loop and acts as a CPL in a closed loop. The V-I characteristic of CPL is depicted in Fig.1.2 as a graph. Fig.1.3 presents the negative incremental aspect of CPL, and it can be observed that the increase/decrease of the current taken by the CPL will make the CPL voltage decrease/increase, respectively. The output power of CPL ignoring system power losses will be identical to the input power.

Thus, CPL's mathematical model can be deduced as follows:

$$P_{in} = P_{out} = constant \tag{1.1}$$

Where P_{in} is input power and P_{out} is output power and are constants.

The mathematical representation of input power with voltage and current can be expressed as :

$$P(v, i) = v.i \tag{1.2}$$

Here, $P(v, i)$ represents the input power, and v and i represents voltage and current, respectively.

The variation in input power is estimated as follows:

$$\delta P(v, i) = v\delta i + i\delta v = 0 \tag{1.3}$$

Thus, the dynamic impedance is calculated as:

$$\delta z = \frac{\delta v}{\delta i} = -\frac{v}{i} \tag{1.4}$$

$$\delta z = -\frac{P}{i^2} \tag{1.5}$$

Hence, mathematically also, it can be checked that the CPL behaves like a negative impedance and has a nonlinear characteristic. Positive impedance has traditionally been associated with the oscillation damping of the LC filter in a circuit and energy dissipation. Negative impedance, on the other hand, causes oscillations to intensify rather than be damped out, introducing instability into the system [30].

When subjected to disturbances, an improperly damped DC microgrid may quickly encounter excessive voltage oscillations. Furthermore, a large CPL may stimulate voltage collapse in a DC microgrid.

1.2.2   Literature review on control design

Due to limited resources, increased economic expenses, and the reduction of dangerous low voltage, research in DC microgrids has expanded in recent years. A DC microgrid connected to DER via an interface with power electronics-based converters can accommodate both traditional and non-traditional loads.

The majority of traditional loads have a constant impedance. On the other hand, non-traditional loads, which fall under the class of CPLs, evolved with the advent of contemporary power electronics-based model. CPLs are nonlinear and don't behave like typical loads.

The CPLs have a characteristic of negative impedance due to nonlinearity, which forces the system towards instability. Because DC microgrids have lower power and energy ratings than large grids, stability difficulties are more prevalent, and appropriate measures are required to make the microgrid practicable. It's difficult to ensure the DC microgrid's stable operation, accurate voltage and current regulation, and system-level management around all operating points caused by the nonlinearity imposed by the CPL.

Also, Power electronics device uses are expanding as technology advances. Various techniques for dealing with this problem using multiple sorts of control are already present in the literature.

The previous literature study based on the existing controller design of DC microgrids connected to distributed generators is sorted and reviewed in this section to acquire a comprehensive review. Control methods to stabilise the DC microgrid can be categorized as linear, nonlinear, or intelligent.

These three categories are described in this section.

1.2.2.1   Linear controller design

This section examines the strategies that strive to exploit the possibilities of employing linear techniques for synthesizing a control law by attempting to achieve maximum stability. Droop control is this category's most often applied control practice for regulating the bus voltage for a DC Microgrid. The voltage at the DC bus is adjusted depending on the power output and input offset[31]. Despite this, droop's application in its most basic version is limited by many drawbacks. The fundamental drawback of the droop controller is that it does not maintain the voltage for the DC-bus constant when the output load power fluctuates. Another constraint is inaccuracy in voltage transfer, leading to decreased current sharing [32].

The other technique employs impedance ratio principles for the systems connected to CPLs, and the system stability is verified using the Nyquist stability theorem[33]. This approach can accomplish the system's desired phase margin and gain margin but may not assure system stability because, in this approach, impedance coincides between the subsystems. Various linear analytical techniques, for example, the Routh–Hurwitz stability criterion, Nyquist plot, and Bode plot are utilised to verify the local stability of the system[34],[35].

These approaches are simple, but the small-signal properties put a limit on their validity region. Linear control design approach can deliver stable operation in the presence of smaller disturbances but are incapable of dealing with instability during the occurrence of big disturbances owing to nonlinear CPL. The characteristics of the linear controller change depending on the types of loads and multiple sources linked to the arrangement. Consequently, control settings cannot be effectively optimised having a large range of operating conditions. As a result, providing global stability via a linear controller of the requisite equilibrium state is impossible. Fig. 1.5 displays the block diagram of the linear control design.

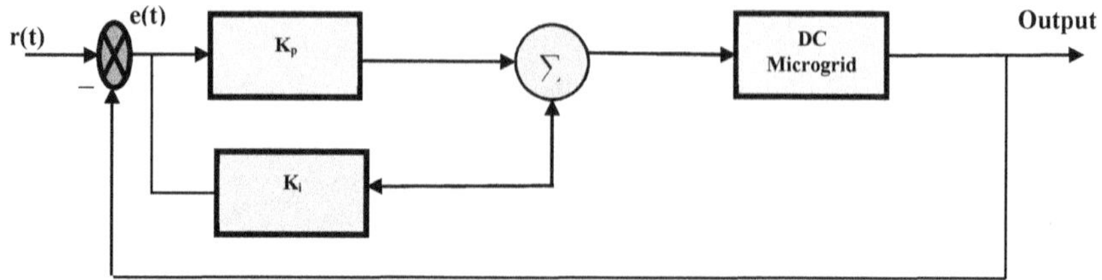

Fig.1.4: Linear PI Control Design

1.2.2.2   Nonlinear Controller Design

Linear controllers are incapable of addressing instability issues during fast transients and considerable fluctuations in load demand. A nonlinear control method is essential to reduce the adverse effects of CPL. Numerous nonlinear control strategies are available in the literature to address control difficulties and other negative impacts of CPL. Sliding-mode method provides a nonlinear design approach in which a control signal is used to drive the system beside the cross-section of its typical characteristic. Sliding-mode operation is a adjustable control design approach.

The control technique is designed to allow trajectories to move beside the limits of the controller structures. Various scientists have proposed this strategy to resolve automation difficulties owing to its rapid dynamical response. The sliding-mode approach is less susceptible to changes in strictures and outside disturbances, resulting in increased system robustness[36],[37]. The sliding mode control strategy can entirely reject matched uncertainties but can't deal with mismatched uncertainty. Another drawback of this approach is that selecting the time-varying sliding surface is time-consuming. Also, the sliding surface should be carefully chosen. Otherwise, the performance of the controller deteriorates. Backstepping control, a recursive technique used to build control laws for feedback using Lyapunov theory, is another alternative[38].

The Backstepping technique is capable of dealing with substantial uncertainty while also ensuring global stability for the system[39],[40]. Using a derivative term in the Backstepping controller is its biggest shortcoming, as it magnifies noise. This method cannot regulate the nonlinearity consequences generated by a steady power load during a disturbance. The H-infinity approach offers increased control with good resilience, allowing the structure to stabilise under undesirable operating conditions such as parameter fluctuations, high disturbance conditions, and model uncertainty[41].

One of the more recent approaches to the H-control design challenge is to tackle an optimization challenge using linear matrix inequality(LMI) limits. The LMI method is employed to govern an effective and robust control design for dynamic structures. The control issue is assembled to form an optimization approach and limitations are expressed in LMI form. The problem of nonlinear control using the H-infinity approach is solved in [42][43] using the inequality methodology and combining the insight of the standard approach. The approach ensures the controller's reliability and effectiveness even though real plant characteristics vary from the assessed ones. This approach delivers a steady result with a robust control design and estimates a reliable solution region. Robust design methodologies deliver satisfactory performance over a given range of plant parameter variations. It produces a fixed gain control that is insensitive to parameter variations. Still, a robust control scheme has the disadvantage of being insensitive to parameter uncertainty.

Nonlinear controllers, as mentioned above, can overcome linear controllers' limitations in sustaining operations over a large operating area. However, several challenges remain unsolved, for example, a dynamic response at transients caused by load and source fluctuations and system analysis with communication delays. Fig. 1.6 displays the block diagram of the nonlinear sliding-mode approach for control design.

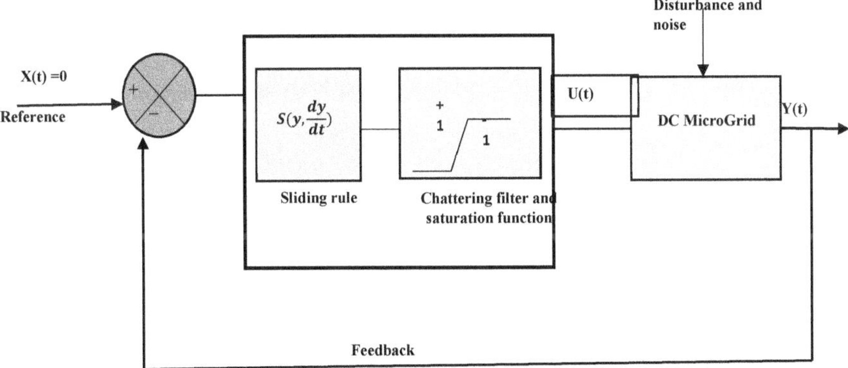

Fig.1.5: Non-Linear Sliding Mode Control Design

1.2.2.3   Intelligent control

Intelligent control techniques employ artificial intelligence techniques like fuzzy-logic, neural networks, and the adaptive neuro-fuzzy inference system (ANFIS). Fuzzy-logic control (FLC) is one of control design's significant research areas. The essential component of a fuzzy logic system that focuses primarily on decision-making is the fuzzy inference system. Fuzzy inference is a technique that evaluates the output vector based on interpreting the values in the reference vector and a set of guidelines.

It involves employing fuzzy logic to map an assigned input value to an output. Subsequently, the mapping offers a foundation from which preferences can be made, or configurations can be identified. Fuzzy inference uses several different components, such as fuzzy logic operators, membership functions, and if-then rules. Fuzzy inference systems are available in two categories, both of which can be used: Mamdani (1977) and Sugeno(1985). Sugeno FIS is more flexible than Mamdani FIS since it allows for more parameters in the output. Also, the Sugeno FIS rule outputs' structure makes them better suited for functional analysis than Mamdani FIS[44]. Out of the numerous fuzzy logic strategies for designing a DC microgrid controller along with several CPLs, the control design established with the Takagi-Sugeno (TS) approach, an evolving noteworthy concept[45],[46]. The nonlinear configuration is broken down into different linear subsystems called fuzzy rules.

A common fuzzy design is created via combining the locally assembled fuzzy-rules. Even though the TS design approach is equally straightforward as the linear technique application, the implementation of fuzzy logic-based control design is substantially competent than linear control methods. The correct training of the fuzzy design is a major difficulty in the model of a fuzzy controller. The model cannot give the desired control if it is not adequately trained. The defuzzification of data is another issue.

Defuzzifying massive amounts of data become a time-consuming job. An alternative approach is a neural network(NN) based control design appropriate for adaptive control. NNs can effectively construct controllers when the mathematical structure of the plant's dynamics is not accessible[47]. The NN is a popular alternative for modelling nonlinear systems and creating general-purpose nonlinear controllers due to its ability to approximate any function. The error of the reference model output and the system output is sent to an adaptive algorithm. This tracking error is reduced by updating the controller parameters.

NNs can also offer a significant degree of fault tolerance because the damage to a few weights need not significantly affect overall performance. Real-time implementations are possible due to NN's large parallel computing structures. Although the closed-loop performance of NN controllers for time-invariant systems with minor tracking errors is guaranteed, the technique for nonlinear control design might be constrained by complex sensor data, noisy training data, and modelling errors[48]. Fuzzy systems can represent extensive linguistic knowledge but lack a mechanism for automatically acquiring or refining such rules.

In contrast, NNs are adaptive systems that can be trained and customised using a collection of samples. NNs can handle new input data by generalising the learned information. Yet, it is quite challenging to extract and comprehend that knowledge[49]. Jang [50] first suggested the ANFIS design in 1993. ANFIS approach creates a fuzzy inference system, which uses the NN learning algorithm to approach a nonlinear system. The ANFIS controller unites the advantages of NN and fuzzy logic. These controllers are more suited to nonlinear systems because of their quick response time and high efficiency. The controller design based on the ANFIS approach uses both the detailed learning skill of neural networks and the robust principles of fuzzy controllers. It uses IF-THEN rules for decision-making to regulate the system[51]. An ANFIS-based design approach is based on applying a sophisticated adaptive network for modelling complex nonlinear systems with a small number of inputs. In this study, a controller is designed using intelligent control methods based on the ANFIS.

The following list includes some of the major advantages behind using the ANFIS algorithm for control design:

- provides robust design for parameter uncertainties
- an additional degree of freedom improves system description
- lumped parameters offer excellent system approximation

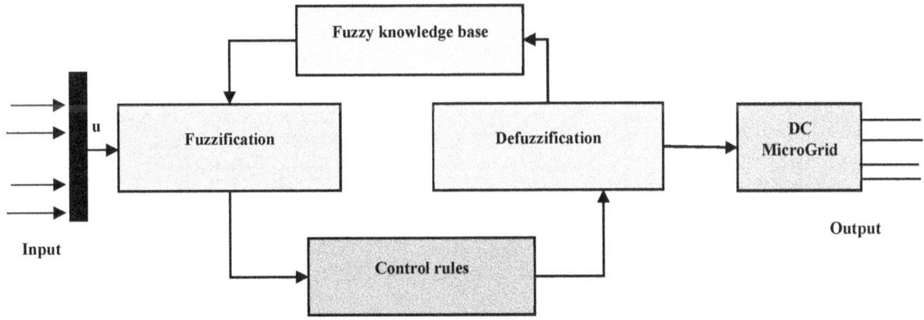

Fig.1.6: Fuzzy-based Intelligent Control Design

Fig. 1.7 displays the fuzzy-based intelligent control design

1.3   Motivation and Objective

Developing a DC microgrid control design and conducting a stability analysis is difficult since the interfacing units are power electronics-based devices with nonlinear dynamics and feed nonlinear loads. The majority of existing power system loads are connected through power electronic devices like converters and inverters, which include an internal control voltage method that can adjust the output voltage. These loads have negative impedance properties and are nonlinear in nature. These loads degrade DC microgrid stability and can result in blackouts or brownouts. Furthermore, the use of power electronics in DC microgrids for integrating renewable energy sources and loads has increased rapidly with the improvement of technology. Thus, with the CPL-based system continuously growing, it is becoming increasingly necessary to stabilize it. An intelligent control procedure is necessary to provide robust and cost-effective control with the aim of stabilising the DC microgrid system to the required degree to solve this critical challenge.

This literature review shows that most control design approaches are based on linearised models of the CPL. However, most of the nonlinear characteristics have been eliminated by linearising the model, which weakens the system's stability. The control design has been researched with several linear and nonlinear design approaches, as can be observed from the reviewed literature. However, several topics, such as transient analysis during frequent load changing and fault conditions for dynamic CPLs, calculating large signal stability criterion, and evaluating it during these conditions, still require further exploration.

Thus, the goals of this dissertation are as follows:

(i)   Designing a nonlinear controller for dynamic CPLs fed to a DC microgrid.

(ii)   Using a computationally intelligent technique to keep the voltage level within the specified limit.

(iii)   Applying stability theory and developing conditions to ensure large signal stability.

(iv)   Optimising the closed-loop performance of a DC microgrid and minimising oscillatory behaviour and the decay rate of the closed-loop DC microgrid.

1.4   Thesis Contributions

This thesis's main contribution is designing a nonlinear controller for a DC microgrid. The control techniques suggested are based on an adaptive robust framework and can manage the nonlinearity caused by the CPLs coupled with the DC microgrid. The proposed design technique can manage both the initial transient response and the difficulties brought on by large load changes and fault conditions.

(a) A robust Control design approach to stabilise DC microgrid with multiple dynamic CPLs :

The increased necessity and usage of ecological renewable energy resources in DC microgrids pose the risk of making the structure unstable, owing to the power electronics-based devices connected to it. These power electronic-based devices have CPL features and are nonlinear. To address the difficulties posed by CPLs in DC microgrids, a robust nonlinear control design approach with Hardy-space matrix values using LMI is adopted. The controller design can successfully stabilise the system during fault events and uncertain variations in load while addressing the issues posed by nonlinear loads. The design approach can guarantee stability by increasing the domain of attraction. As a result, the method can address the issues with multiple operating points brought on by nonlinear loads. The design approach's efficacy is verified with simulation done in MATLAB/SIMULINK

(b) Adaptive Framework with Intelligent Control to stabilise DC microgrid with multiple dynamic CPLs :

An adaptive control design technique is put out in this study with stability proofs. The adaptive controllers are derived for uncertain nonlinear loads. The control design approach utilises ANFIS-based control design, which is essentially adaptive. The uncertain load power is estimated using the observer-based estimation method. The CKF is chosen for state estimation of load power; the estimated power is used in the control design. The CKF's convergence analysis is also provided to prove the efficacy of the observer.

(c) Adaptive Robust Framework with Intelligent Control to stabilise DC microgrid with multiple dynamic CPLs :

A combination of the adaptive and robust design has been suggested for the controller design to lessen the effects of nonlinear CPLs connected to the system. A suitable stability analysis based on the Lyapunov theorem is also provided. This work proposes a combined strategy using an adaptive and robust design approach for controlling DC microgrids coupled with various dynamic CPLs. The suggested controller is competent in handling the variations in bus voltage under fault conditions and large load changes effectively.

1.5   Outline of the Thesis

There are seven chapters in this thesis. This thesis's contents are arranged as follows :

Chapter 1 provides a brief introduction and a list of prior contributions in this area. The rising use of DC microgrids in the power sector, as well as the challenges in maintaining control and a suitable stability analysis for DC microgrids, are covered in this chapter. This serves as the driving force behind the writing of this thesis. The research objectives for DC microgrid control design are explained. A summary of key contributions has been highlighted.

The approach for mathematical modelling of uncertain nonlinear systems, system error identification, and stability analysis that are employed throughout the thesis are described in Chapter 2. Large signal stability criteria using the Lyapunov theorem, proof of stability theorem, and approach for estimating the domain of attraction are discussed in this chapter. Observer design for nonlinear system state estimation with uncertainty is also covered.

In Chapter 3, the overall state-space model of the n-bus DC microgrid is explored, including the robust control framework for nonlinear CPLs coupled to DC microgrids. The control architecture for CPLS connected to a DC microgrid is proposed using a Hardy space-valued matrix function. Investigations are also conducted for the stability analysis of the framework indicated above based on Lyapunov stability criteria. Lyapunov stability is verified using the LMI approach. Finally, simulation findings are applied to confirm the usefulness of the suggested concept.

Chapter 4 discusses the adaptive framework with intelligent control for CPLs connected to DC Microgrid. It proposes an ANFIS-based controller design that regulates the DC voltage for multiple dynamic CPLs linked to the DC microgrid. This chapter uses a CKF to assess dynamic CPL's power. The evaluated power is then used in an ANFIS controller. Convergence analysis for CKF is also provided, along with the stability evaluation of the suggested control design. Simulation findings are also discussed to confirm the efficacy of the suggested method.

Chapter 5 explores the adaptive robust framework for DC microgrid control design. The controller is designed using an ANFIS-based technique, and an observer based on HCKF is proposed for state estimation. A Lyapunov-based stability analysis is used to validate the suggested controllers' stability for large signal variations. Finally, simulation findings are presented to validate the effectiveness of the proposed controller design.

Chapter 6 provides the experimental findings using hardware in the loop simulation.

The conclusion of this thesis is provided in Chapter 7, along with suggestions for future investigation.

CHAPTER2

FUNDAMENTALS OF MICROGRID STABILITY AND CONTROL:

2.1 Modelling of Uncertain nonlinear systems:

The main causes of uncertainty are defects in the system, sensitivity to initial conditions, the uncertainty of system parameters and external loads, and interaction between load and system. These types of uncertainties, when combined with nonlinearities in the system, can significantly influence the system's response, mainly in dynamic conditions. Consequently, it's essential to explore their impact on the stability of a nonlinear structure[52].

Considering the following nonlinear approach with uncertainties:

$$\begin{cases} \dot{x}(t) = g(x,t) + \Delta g(x,t) + O(x(t))u(t) \\ \qquad x(0) = x_0, t \geq 0 \end{cases} \tag{2.1}$$

Where $x(t) \in R^n, t \geq 0$ is the vector for the state, $u(t) \in R^m, t > 0$ is the control input. $g :$ $R^n \to R^n$ and satisfies $g(0) = 0, \Delta g: R^n \to R^n$ and satisfies $\Delta g(0) = 0$ and $O: R^n \to R^{nxm}$. Here, we assume that $g(.)$ and $\Delta g(.)$ are uncertain with $\Delta g(.)$belonging to the uncertain set B given by :

$$B = \{\Delta g: R^n \to R^n: \Delta g(x) = O_\delta(x)\delta(h_\delta(x)), x \in R^n, \delta(.) \in \Delta\} \tag{2.2}$$

Here, Δ assures

$\Delta = \{\delta: R^{i\delta} \to R^{j\delta}: \delta(0) = 0, \quad$ and $\quad \delta^T(y)\delta(y) \leq P^T(y)P(y), y \in R^{i\delta}$ .Also, $O_\delta: R^n \to$ $R^{nxm_\delta}, h_\delta: R^n \to R^{nxi_\delta}$ assuring $\delta(0) = 0$ are fixed functions indicating uncertainty form. δ is an uncertain function.

The control $u(.)$ is restricted to the set of controls that consist of considerable functions such that $u(t) \in R^m, t \geq 0$. Additionally, for the uncertain nonlinear system O , it is assumed that the essential properties for the existence and uniqueness of solutions are fulfilled. $g(.), \Delta g(.), O(.), u(.)$ comply with enough stability conditions such that the equation has a single solution after $t \geq 0$. Here i and j present the ith and jth term of δ (uncertain function).

2.1.1 System error identification:

The control design aims to guarantee that the desired time-varying trajectory will be tracked irrespective of structured and unstructured uncertainty in the dynamic configuration for the specified system. The idea is that the controller design is such that it guarantees the closed-loop system performance, and in addition, the outputs $y_1, y_2, \ldots y_n$ track the reference inputs $x_1, x_2, \ldots x_n$ respectively[53]. The system dynamics are combined with the controller state vector to incorporate the tracking capabilities.

Let the tracking error, $e(t)$, be identified as:

$$e(t) = r(t) - y(t) \tag{2.3}$$

Where $e(t), r(t), y(t) \in R^n$. $r(t)$ is reference input, and $y(t)$ denotes the measured output. State dynamics $\dot{x}_n \in R^n$ is given by

$$\dot{x}_n = e(t) \tag{2.4}$$

The dynamic system under consideration has the following specifications:

$$\begin{cases} \dot{x}(t) = f(x(t) + Bu(t) \\ \quad y(t) = C(x(t)) \end{cases} \tag{2.5}$$

Where $x(.) \in R^n$ presents State vector, $u(.) \in R^m$ presents control vector, $y(.) \in R^p$ presents output vector. Also, $f(x)$ is continuously differentiable with respect to x and stated as :

$$f(x) = Ax + \emptyset(x) \tag{2.6}$$

Where $\emptyset(x)$ is a function that obeys $\emptyset(0) = 0$ without loss of generality.

Combining (2.4) with (2.5), the system with a closed loop can be stated as:

$$\begin{bmatrix} \dot{x} \\ \dot{x}_n \end{bmatrix} = \begin{bmatrix} A & 0 \\ -C & 0 \end{bmatrix} \begin{bmatrix} x \\ x_n \end{bmatrix} + \begin{bmatrix} B \\ 0 \end{bmatrix} \hat{u} + \begin{bmatrix} 0 \\ r(t) \end{bmatrix} + \begin{bmatrix} \emptyset(x) \\ 0 \end{bmatrix} \tag{2.7}$$

Where control signal $\hat{u}$ can be calculated as :

$$\hat{u} = \begin{bmatrix} \bar{k} & k_n \end{bmatrix} \begin{bmatrix} x \\ x_n \end{bmatrix} = k_q L \tag{2.8}$$

Here $\bar{k} \in R^{m \times n}$, $k_n \in R^{n \times p}$, $k_q \in R^{n \times (m+p)}$. Rewriting the (2.7) as :

$$\dot{L} = A_q + B_q k_q + R(t) + \emptyset(L) \tag{2.9}$$

Where $\dot{L}$ is a general state of a system, A_q, B_q and k_q are state matrix, input matrix, and control gain matrix for qth vector, respectively, and function $\emptyset(L)$ follows:

$\|\emptyset(L)\| \leq \|\gamma(L)\|$ and γ is the Lipschitz constant for $\emptyset(L)$.

Below are the explanations of the notations:

$$L = \begin{bmatrix} x \\ x_n \end{bmatrix}, A_q = \begin{bmatrix} A & 0 \\ -C & 0 \end{bmatrix}, B_q = \begin{bmatrix} B \\ 0 \end{bmatrix}, R(t) = \begin{bmatrix} 0 \\ r(t) \end{bmatrix}, \emptyset(L) = \begin{bmatrix} \emptyset(x) \\ 0 \end{bmatrix} \tag{2.10}$$

Including $L \in R^{(m+n)}$, $A_q \in R^{(m+n)(m+n)}$, $B_q \in R^{m(n+p)}$, $\emptyset(L) \in R^{m+p}$ and $R(t) \in R^{n+p}$.

Here the objective is tracking the reference signal $r(t)$ by means of an output vector $y(t)$ for ensuring the stability of the control signal.

2.2 Stability Analysis for the uncertain nonlinear system:

2.2.1   The approach used for stability analysis :

There are several techniques to investigate nonlinear system stability. Out of these, Eigenvalue determination and the Lyapunov method are the most prevalent techniques for assessing nonlinear stability. The first technique is to locate the eigenvalues estimated from the state matrix after linearising the model. Another approach is to assess the asymptotical stability area (domain of attraction) utilizing the Lyapunov method. Simulating the mathematically applied model is another alternative for conducting the analysis.

The state matrix's eigenvalues can be applied to evaluate the global asymptotical stability of a dynamic structure. The method is asymptotically stable when all of the eigenvalues contain real but negative components. Since the model is linearised, the stability analysis done by this method can only guarantee asymptotic stability near the equilibrium point. It is crucial to check the stability of a nonlinear system for large signal deviations. The Lyapunov method is best suited for analysing the large-signal stability of a nonlinear model[54].

Applying the Lyapunov theorem requires access to the Lyapunov function. The function can be formulated using several different methods. In this study, the biggest estimate of the Domain of Attraction(DA) of nonlinear systems is obtained using the Hardy space-infinity approach[55].

Consider the nonlinear model with n distinct nonlinearities. It is represented as follows:

$$\dot{x}(t) = \sum_{i=1}^{n} w_i(x)\,(A_i(x(t) + B_i u(t)) \tag{2.11}$$

$$y(t) = \sum_{i=1}^{n} w_i(x) H_i x(t) \tag{2.12}$$

In the above equation, each local linear model is given a normalised weight $wi(x)$ to get the nonlinear model. $Ai, Bi,$ and Hi represent the constant matrices.

The following conditions must be met for a nonlinear model to be stable:

$$\begin{cases} P = P^T > 0 \\ A_i^T P + PA_i < 0, \forall i = 1, \dots \dots \dots, 2^n \end{cases} \qquad (2.13)$$

Therefore, it is adequate, but not necessary, to demonstrate the nonlinear model stability if a general positive definite matrix P exists which fulfils the Lyapunov inequality of all the 2^n local models. The Lyapunov function, in this case, is:

$$V(x) = x^T P x \qquad (2.14)$$

Consequently, an estimate of the DA is the region in which all of the $2k + 1$ inequalities in (2.13) hold.

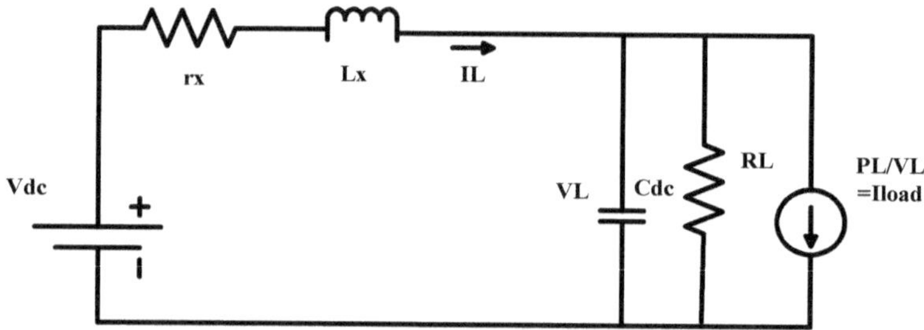

Fig.2.1: Equivalent Circuit of DC Microgrid using Single CPL

Only the CPL contributes to the nonlinearity in the DC microgrid considered here. The nonlinear system configuration used for the DC microgrid with a single CPL is displayed in Fig.2.1 . The model displayed in Fig.2.1 can be expressed in the state equation form as:

$$\begin{cases} \dot{x}_1 = -\dfrac{r_x}{L_x} I_L - \dfrac{1}{L_x} V_L + \dfrac{1}{L_x} V_{dc} \\ \dot{x}_2 = \dfrac{1}{C_{dc}} I_L - \dfrac{1}{C_{dc}} \dfrac{P_L}{V_L} \end{cases} \qquad (2.15)$$

Here x_1 presents inductor current and x_2 presents capacitor voltage. C_{dc} is the output voltage. I_L and V_L are load current and load voltage, respectively. V_{dc} denotes the source voltage. r_x , L_x are the line parameters representing resistance and inductance, respectively.

The current and voltage for the Bus around the equilibrium points are considered as I_e and V_e for a given load power P_L. The stability for the large signal deviation around the equilibrium point (I_e, V_e) can be analysed by shifting the state vector from its equilibrium point as:

$$\begin{cases} x_1 = i_L - i_e \\ x_2 = V_L - V_e \end{cases} \tag{2.16}$$

The model (2.15) can then be expressed as:

$$\begin{bmatrix} \dot{x}_1 \\ \dot{x}_2 \end{bmatrix} = \begin{bmatrix} -\dfrac{r_x}{L_x} & -\dfrac{1}{L_x} \\ \dfrac{1}{C_{dc}} & f(x) \end{bmatrix} \begin{bmatrix} x_1 \\ x_2 \end{bmatrix} \tag{2.17}$$

Here $f(x)$ shows the nonlinearity contributed by the CPL and can be expressed as:

$$f(x) = \frac{P_L}{C_{dc}V_e(x_2 + V_e)} \tag{2.18}$$

Solution of (2.15) provides the estimate of the region of the DA, as stated in [56]. The MATLAB simulation and hardware experiments can verify this solution in chapter 6.

2.2.2 Large signal stability analysis using Lyapunov Theorem:

The definition of the Lyapunov function serves as the foundation for the stability analysis for large signal deviation, and it asserts that :

for the system $\dot{x} = f(x)$, A function $V(x): Rn \rightarrow R$ is a Lyapunov function in $\vartheta(0, \rho)$ if it meets the following criteria

(1) $V(x)$ is positive definite in $\vartheta(0, \rho)$
(2) $V(x)$ holds continuous first-order partial derivatives with respect to x
(3) $V(x) = \langle \nabla V, f(s) \rangle \leq 0 \ \forall x \in \vartheta(0, \rho), \langle \nabla V, f(s) \rangle = \langle \nabla V, \dot{x}(t) \rangle$

According to Lyapunov, the following conditions must be met to achieve asymptotic stability[57]:

2.2.3 Asymptotic stability:

In the context of Lyapunov, the zero state is asymptotically stable. It can be proved provided a Lyapunov function $V(x)$ exists in the vicinity of the starting point $\vartheta(0, \rho)$ in a way that $dV(x)/dt < 0$ for all $x \in \vartheta(0, \rho), x \neq 0$. In other words, if a function V exists such that

- V is a positive definite
- $V(p) < 0$ for each $p \neq 0, \dot{V}(0) = 0$

then, as $t \rightarrow \infty$ every trajectory of $\dot{x} = f(x)$ approaches to zero, making the system globally asymptotically stable.

2.2.4 Proof of stability theorem:

Assume that the $x(t)$ trajectory does not converge to zero. Let's say $V(x(t))$ converges to ε, as $t \to \infty$, and it is nonnegative and decreasing. We must have $\varepsilon > 0$ because $x(t)$ does not converge to 0; hence for any t, $\varepsilon \leq V(x(t)) \leq V(x(0))$.

$C = \{p \mid \varepsilon \leq V(p) \leq V(x(0))\}$ is closed and bounded, hence compact.

So in other words, $\dot{V}$ reaches its supremum on C, i.e. $sup_{z \in C} \dot{V} = -a < 0$, where $\dot{V}$ assumed to be continuous.

As $\dot{V}(x(t)) \leq -a$ for all t, we have $V(x(T) = V\big(x(0)\big) + \int_0^T \dot{V}\big(x(T)\big)dt \leq V(x(0) - aT$, Where for $T > V(x(0)/a$ indicates $V(x(0) < 0$ an inconsistency.

Hence, every trajectory $x(t)$ converges to 0, which means $\dot{x} = f(x)$ is globally asymptotically stable.

2.2.5 The domain of asymptotic stability:

Let h be a real number that is positive, and let $V(x)$ be a Lyapunov function. This will ensure that the open set $D = \{x : V(x) < h\}$ to be bounded and let $dV(x)/dt < 0$ for all $x \in D, x \neq 0$. Following that, all paths originating from a particular position in the set D approach zero asymptotically. A nonlinear system's asymptotic stability domain is referred to as a region of the state space where each trajectory starts and finally converges to equilibrium[58].

2.2.6 Estimate for the domain of attraction(DA) :

Consider a system $\dot{x} = f(x)$ and $V(x)$ represents a Lyapunov function at the equilibrium point $x = 0$. Consider that $dV(x)/dt$ is a negative definite in the specified domain:

$$DA(0) = \{x : V(x) = c, c > 0\} \tag{2.19}$$

Following that, as time tends to infinity, every trajectory started in region $DA(0)$ tends to $x = 0$. According to (2.19), the estimation of DA (0) can be improved if the level set value c has a larger value. The largest degree set of the Lyapunov function, which is even an evaluation of DA(0), can be calculated by resolving a problem with the following quasi-optimization formulation: Quasi optimisation formulations are a class of optimization techniques that are used for non-linear functions. These techniques are applied to obtain the global optimum for a function $f(x)$.

$$max_{c,x} \ c$$

$$s.t. \{x \text{ belong to level set } V(x) - c = 0\} \tag{2.20}$$

The purpose of the problem in (2.20) is to get the largest level set of $V(x)$ that is entirely covered in the area of negative definiteness of $dV(x)/dt$. This can be achieved by answering the subsequent minimization problem to provide global optimisation [59].

$$min_{c,x} \; c$$

$$s.t. V(x) - c = 0 \tag{2.21}$$

$$\frac{dV(x)}{dt} = 0$$

$$c > 0$$

The goal of the problem (2.21) is to identify the smallest level set of function $V(x)$ that may be included within the level set $dV(x)/dt = 0$. A specific position in the state space that explains a link between the level sets $V(x) = c$ and $dV(x)/dt = 0$ represents the ideal solution. It should be emphasised that since model (2.21) is nonlinear, multiple local solutions may exist. Global optimality must be guaranteed for the evaluation to approach the correct value.

The DC microgrid becomes a nonlinear system with the addition of the CPL, and the stability analysis can be performed with the help of the Lyapunov theorem for nonlinear systems. Stability tests for small-signal variations can only ensure stable control close to an equilibrium point. Conducting large-signal stability tests to get an asymptotically stable operation in the DA region is necessary.

2.3 State Observer design for Nonlinear System with uncertainties

Nonlinear control design requires accurately estimating a dynamical system's current state. A widely accepted approach to this challenge is to install sensors in the physical system or to create an observer. Due to financial and physical limitations, sensors' quantity and quality are frequently constrained in practical situations. Observers are necessary for a control architecture to deal with the imperfect and partial information required by the sensors and produce a trustworthy assessment of the status of the entire system.

The challenge of observer design arises whenever a system approach requires some internal knowledge from external measures. Internal information is necessary to identify the technique, detect errors, and maintain it. As shown in fig.2.2, all those objectives are essential while attempting to maintain system control. As a reason, the observer, the problem stands at the centre of the general control problem.

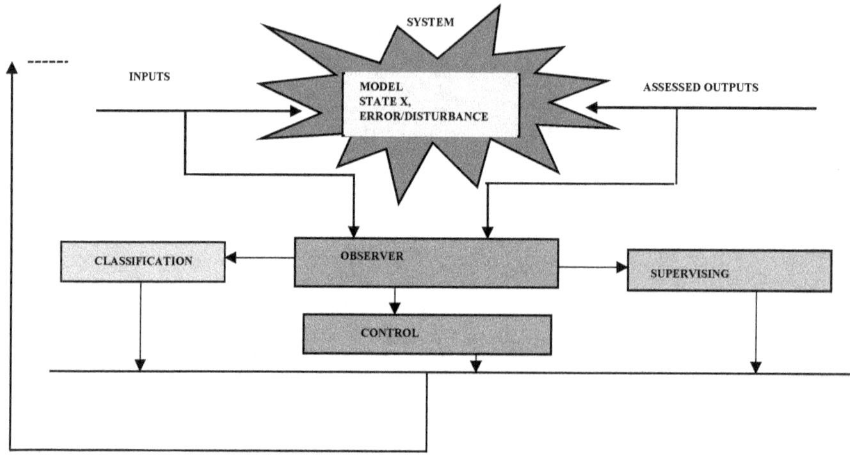

Fig.2.2: Block Diagram for Observer

Preliminary observer designs for non-linear systems were introduced in [60] and [61] and were developed in [62-64] by using a state transformation to produce uniform dynamic behaviour for errors. In [65], combining a local and a global observer is suggested to assess the state of a general continuous nonlinear system. A dynamic equation of nonlinear observers is presented in this chapter that can handle an uncertain nonlinear system.

The suggested theory for determining gain for an observer is established on the Lyapunov function's convergence within the Linear Matrix inequality (LMI) optimization theory framework and can be easily resolved using MATLAB.

2.3.1 Nonlinear Observer Construction:

Since the model is frequently a state-space representation, it will be assumed that all of the information that has to be reconstructed is produced by state variables. In addition, one can either formulate the task as an optimization problem or construct an actual dynamical system, the state of which should yield an assessment of the real state of the model being studied.

Consider the uncertain nonlinear continuous system shown below:

$$\begin{cases} \dot{x} = f(x(t), u(t)) \\ y(t) = g(x(t), u(t)) \end{cases} \tag{2.22}$$

Where $x \in R^n$ is the state vector, $u \in R^q$ is control vector, $y \in R^p$ is output vector, and f, g, are nonlinear vector functions of dimensions n, p respectively.

The structure of an observer for a nonlinear dynamic system is as follows:

$$\begin{cases} \dot{\hat{x}}(t) = f\big(\hat{x}(t), u(t)\big) + k\big(y(t) - \hat{y}(t)\big) \\ \qquad \hat{y}(t) = g(\hat{x}(t), u(t)) \end{cases} \tag{2.23}$$

The nonlinear observer is therefore represented as:

$$\dot{\hat{x}}(t) = f\big(\hat{x}(t), u(t)\big) + k(y(t) - g(\hat{x}(t), u(t))) \tag{2.24}$$

The observer gain K represents a nonlinear matrix function that depends on x and u. It must be chosen so that the observation error, $e(t)$, approaches zero. The observation error dynamics are calculated by:

$$\dot{e}(t) = \dot{x}(t) - \dot{\hat{x}}(t) = f\big(x(t), u(t)\big) - f\big(\hat{x}(t), u(t)\big) + k(g\big(x(t), u(t)\big) - g(\hat{x}(t), u(t))) -$$
$$\text{---}(2.25)$$

The solution of this equation is an error approaching zero, indicating that the designed observer may have $e = 0$ at a steady state. To ensure the observer and error dynamics' asymptotical stability, the gain K must be selected carefully. The challenge is to create a dynamic system which satisfies the equation given below:

$$\lim_{t \to \infty} |\hat{x}(t) - x(t)| = 0 \tag{2.26}$$

2.4 Summary and Findings:

The small signal stability analysis can ensure stable working close to the operating point only. Large signal stability tests are necessary to achieve asymptotic stability near the DA. The control design established on large signal stability criteria needs a proper mathematical model. This chapter discusses a detailed mathematical model for the DC microgrid applied for this research. This chapter also discusses the approach for large signal stability criteria and proof of that analysis.

Since the DC microgrid is uncertain and linked to nonlinear loads, a detailed approach for system error identification is provided in this chapter. Nonlinear observers are required to estimate the uncertainties associated with nonlinear load power. The observer design approach, which is used in this research, is also discussed in this chapter.

Finally, an estimate of the domain of attraction is added to provide proof of stability for the suggested controller design.

CHAPTER 3

ROBUST CONTROL FRAMEWORK:

3. Robust Control Framework for CPL-connected DC Microgrid:

Modern DC microgrid control approaches have trouble generating successful and reliable results. Heuristic and intelligent control systems cannot guarantee simultaneous robust stability and function for various disturbances and uncertainties. In addition, any electrical network's stability is essential. Here stability indicates a system's capability to sustain generation-consumption power stability in the face of a variety of threats, including random power requests or injections, changing physical circuits, and so on. An islanded microgrid's stability is much more at risk because it lacks any grid backup and must resolve any potential power imbalance using just local resources.

Implementing a robust design that enhances the system's transient performance and minimizes the steady-state tracking error is necessary. The sliding mode controller (SMC) is the one approach that can be used for robust design because of its stability, robustness, and high compatibility with the switching converters.

The SMC method must develop a control law conditional on a specific sliding surface to provide asymptotic convergence. However, It is difficult for this technique to achieve the equilibrium point since it cannot be reached in a finite amount of time. Also, this technique has a chattering issue, which lessens its effectiveness for a robust design.

Another strategy is the fuzzy PID controller, which combines fuzzy logic and a traditional PID controller. A system's performance can be improved using a fuzzy-based PID controller without a mathematical model. The controller, however, is inappropriate for nonlinear plants since it might not guarantee the necessary performance under varying operating conditions.

Though the linear robust control approaches provide effective control synthesis methodologies for dynamical systems by considering physical constraints and uncertainties, the approach offers complex designs for controllers with orders that are not smaller than those of controlled systems.

Nonlinear robust control techniques, on the other hand, effectively meet this goal due to the potential of uncertainty formulation in the control synthesis procedure. The approach of a nonlinear hardy space controller design for the off-grid operation of a microgrid is presented in this chapter. The suggested controller ensures stable operation under a variety of loads and uncertainties. The suggested controller design, which uses a linear matrix inequality approach, satisfies the Lyapunov stability condition. In addition, it offers a closed-loop H-norm with good tracking accuracy.

The proposed approach, based on the Hardy space nonlinear control principle, solves the problem of microgrid stability for multiple constant power demands and external disturbances. This chapter explores the stability conditions for a DC microgrid with the proposed nonlinear controller design and multiple constant power loads. A robust stability framework with sufficient conditions is presented to determine the DC microgrid stability. Each stability condition involves a convex optimization problem, the feasibility of which indicates the system's stability.

3.1 Mathematical model of a DC Microgrid :

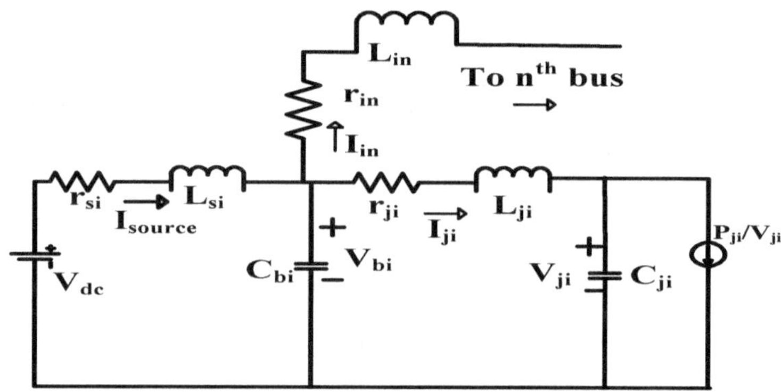

Fig.3.1(a): DC microgrid diagram connected to an ith bus

As indicated in Fig. 3.1(a), we explore a DC microgrid connected to n buses in this research. Each component of the edge set E signifies a transmission line linking two buses, and each member of the vertex set V denotes a DC microgrid bus, forming an undirected graph $G = (V, E)$. The ith bus is supposed to be an adjacent bus to the jth bus if $(i, j) \in E$. Let N_i contain the indices of all the buses that are adjacent to the ith bus.

Assume that $Z \in R^{n \times n}$ is the adjacency matrix. It is assumed for ease that each bus has a single controllable voltage source, and one CPL. Fig. 3.1(a) illustrates the corresponding circuit model of the ith bus. In this part, a state-space dynamic model is developed using ith bus before moving on to the microgrid's overall dynamic model. It's important to note that the modelling approach may easily be used for DC microgrids with a variety of sources or loads on some buses or none at all.

3.1.1 State Space model for ith bus of DC microgrid :

- **Constant Power load(CPL):**

The constant power load can be represented as load power divided by load voltage and is modelled as a current sink. An equivalent RLC filter links the CPL with the microgrid. Let v_{ji} and i_{ji} represent the filter's terminal CPL voltage and current injection, respectively. In parallel with the current source, the voltage of the capacitor across it has the same voltage v_{ji} as the CPL.

The current i_{ji} A resistor and inductor are fed to the load via the DC bus capacitor. Hence, the CPL has two state variables v_{ji} and i_{ji}.

The state variables' dynamics are provided by:

$$\begin{cases} L_{ji}\dfrac{di_{ji}}{dt} = v_{bi} - r_{ji}i_{ji} - v_{ji} \\ \quad C_{ji}\dfrac{dv_{ji}}{dt} = i_{ji} - \dfrac{P_{ji}}{v_{ji}} \end{cases} \tag{3.1}$$

Here P_{ji} represents the CPL power, and the CPL current is represented by $i_{cpl} = \dfrac{P_{ji}}{v_{ji}}$. L_{ji} , C_{ji}, and r_{ji} represents the load inductance, capacitance, and resistance, respectively. The load power is supposed to be uncertain, and the power value lies in between $\left[\underline{P_{ji}} \quad \overline{P_{ji}} \right]$, where $\overline{P_{ji}} \geq P_{ji} \geq \underline{P_{ji}}$. Also,

$$i_{cpl} = \dfrac{P_{ji}}{v_{ji}} \text{ , or } v_{ji} = \dfrac{P_{ji}}{i_{cpl}} \text{ or } \dfrac{\delta v_{ji}}{\delta t} = -\dfrac{P_{ji}}{i^2{}_{cpl}} \tag{3.2}$$

This equation shows that due to negative incremental impedance, constant power adds nonlinearity to the system. Also, the negative term shows that any minor variation in voltage can introduce an adverse current variation while preserving the constant power. This implies if the voltage v_{ji} around the load declines owing to any disruption, the current i_{cpl} will attempt to maintain the power at constant level.

- **Source :**

Assuming each bus has a single controllable voltage source and V_{dc} and I_{source} represents the voltage at the terminal and current output, respectively, on ith bus. The microgrid is linked to the source with the resistance and inductance. The state equation of I_{source} is presented as :

$$L_{si}\frac{dI_{source}}{dt} = V_{dc} - r_{si}I_{source} - v_{bi} \tag{3.3}$$

Here r_{si} and L_{si} are source resistance and inductance, respectively. v_{bi} denotes the voltage of ith bus.

- **DC link capacitor :**

The voltage around the DC bus, v_{bi}, serves as the state variable for the capacitor connected to the DC link, and the state equation is expressed as:

$$\begin{cases} C_{bi}\frac{dv_{bi}}{dt} = I_{source} - i_{ji} - \sum_{i \in N_i} i_{in} \\ \qquad i_{in} = \frac{v_{bi}-v_{bn}}{r_{in}} \end{cases} \tag{3.4}$$

where N_i is the set of indices of all the kth bus's surrounding buses. C_{bi} and r_{in} are the ith bus's DC link capacitance and transmission line's equivalent resistance linking the ith and nth buses, respectively.

3.1.2 Overall state space model for n-bus DC microgrid:

The source, CPL, and DC link capacitor individual models for each bus can be combined to create the total configuration of the DC microgrid. Let i_x , and $v_x \in R^n$ and represents all the currents by sources and CPLs, voltages across DC buses, and CPLs, while p represents the load power. Also, the state vector $x \in R^n$ and the nonlinear function $f(\cdot,\cdot): R^n \times R^{4n} \to R^n$ is defined as :

$$\begin{cases} x = \left[i_x^{\ T}, v_x^{\ T}\right]^T \\ f(p,x) = \left[-\frac{P1}{v_{i1}}, \frac{P2}{v_{i2}}, \dots\dots\dots\dots\dots, \frac{Pn}{v_{in}}\right] \end{cases} \tag{3.5}$$

Where, $i_x = [i_{s1},..,i_{sn},i_{i1},\dots\dots\dots i_{in}]^T$, $v_x = [v_{b1}, v_{b2} \dots\dots\dots v_{bn}, v_{i1}, \dots\dots v_{in}]$,

$p = [p_1, p_2 \dots\dots p_n]$. The nonlinear function $f(p,n)$ contains the incremental nonlinear terms in the equation, and the vector x represents all of the component's current and voltage states. The overall microgrid state equation can be presented as:

$$\begin{cases} \dot{x} = Ax + Bu + f(p,x) \\ \qquad z = Cx \end{cases} \tag{3.6}$$

Here x and z are the state input and output vectors, respectively, and u is the control input matrix. The matrix A denotes the system state matrix, B denotes the input state matrix, and C denotes the output state matrix. The following are the overall dynamic equations of a DC microgrid displayed in Fig. 3.1:

$$\begin{cases} \dot{x}_1 = -\frac{r_x}{L_x}x_1 - \frac{1}{L_x}x_2 + \frac{1}{L_x}V_{dc} \\ \quad \dot{x}_2 = \frac{1}{C_x}x_1 - \frac{1}{C_x}f(x_2) \end{cases} \tag{3.7}$$

Where $x_1 \; and \; x_2$ represents the state vector for the inductor current and capacitor voltage, respectively, whereas $f(x_2) = \frac{P}{x_2}$ represents CPL. Here V_{dc} is the source voltage and r_x characterises the source resistance.

3.2 Control challenges in problem formulation:

A DC microgrid is represented as a space state model in (3.6), which includes all of the generating sources, CPLs, and buses connected between sources and loads. As displayed in Fig.3.1(a), the CPL in a DC microgrid acts like a current sink. The current value is calculated by dividing the load power with the load voltage. These CPLs have negative incremental impedance $v - i$ characteristics due to their nonlinearity, which could lead to system instability.

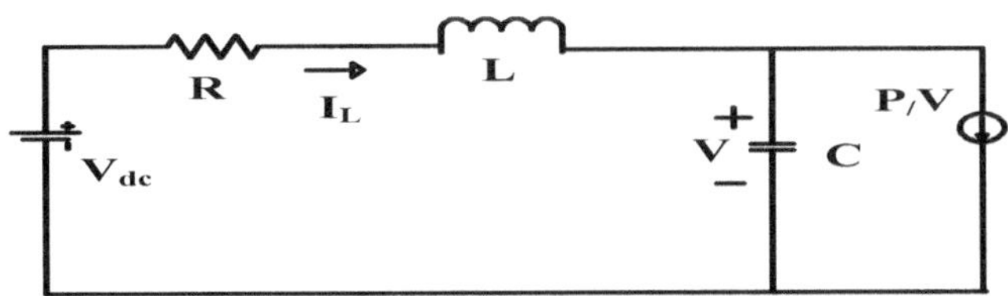

Fig.3.1(b): Equivalent circuit of the single bus closed loop system

The state-space equation of the equivalent circuit given in 3.1(b) can be presented as:

$$\begin{cases} \frac{di_L}{dt} = \frac{1}{L}(V_{dc} - Ri_L - V) \\ \quad \frac{dV}{dt} = \frac{1}{C}\left(i_L - \frac{P}{V}\right) \end{cases} \tag{3.8a}$$

The equilibrium points for bus voltage and current for 3.1(b) are calculated by solving the equation given in (3.8a):

$$\begin{cases} V_{eq} = \dfrac{V_{dc} \pm \sqrt{(V_{dc}^2 - 4Pr_x)}}{2} \\ I_{eq} = P/V_{eq} \end{cases} \qquad (3.8b)$$

If Pr_x is small in (3.8b), the system's equilibrium points will be real and positive. If Pr_x is greater than V_{dc}^2, the equilibrium point is complex. In this situation, no point of operation is physically possible. As a result, to prevent complex equilibrium points, the following prerequisite must be satisfied:

$$r_x < \frac{V_{dc}^2}{4P} \qquad (3.9)$$

The source resistance is limited by this constraint, which is determined only by the source power and the operating voltage. The system becomes unstable when the condition indicated by equation (3.9) is violated. To keep the operation stable, safety precautions must be applied. After using a safety factor of 2, the equation (3.9) can be rewritten as:

$$r_x < \frac{V_{dc}^2}{8P} \ \text{ or } P < \frac{V_{dc}^2}{8r_x} \qquad (3.10)$$

Equation (3.10) is changed to the following form when the source resistance is considered as 5% of the load resistance:

$$P < \frac{V_{dc}^2}{8 * 0.05 r_x} < 2.5 * P_{rated} \qquad (3.11)$$

The condition constrains the CPL, system voltage, and source resistance in equation (3.11), and due to these limitations, equilibrium points for large transients might not be possible.

This condition does not explain the system's overall stability; it merely provides support for local stability [66]. The local stability requirement might not ensure the system's secure operation because of the nonlinearity that the model's function $f(p, x)$ introduces.

These nonlinearities are crucial to stability analysis during large signal variations and cannot be overlooked. The situation also doesn't give any details about the region near the operating point of DC bus voltage where the system can be stabilised.

3.3 Framework for Robust Stability

3.3.1 Robust Stability Framework with multiple CPL

The stability analysis is essential for DC microgrid applications since CPLs can create instability. The undefined power from CPL P is supposed to be confined in a polytopic set S in this study and described as:

$$S = \left\{ s : s_k \in \left[\underline{s_k}, \overline{s_k} \right], k = 1, \cdots, n \right\} \tag{3.12}$$

The challenge is to analyse the stability of system (5) with an unknown $p \in S$ since the equilibria of the nonlinear system is challenging to get. If the model is connected to n number of CPLs, the equilibria must be obtained by solving n quadratic equations.

There are 2^n solutions for the n number of quadratic equations, and determining the real equilibrium to which the system will converge is very difficult. Additionally, as p varies, the equations' solution alters. The set of feasible equilibria of a general DC microgrid cannot be described with an unpredictable CPL power p. The stability is established for the entire domain of attraction(DA) because there isn't a fixed equilibrium. Also, the asymptotic and exponential stability become global when the stability is calculated around the DA.

At the equilibrium point $x = 0$, the DA is expressed as:

$$DA(0) = \{x_0 \in R^n : \lim_{t \to \infty} x(t, x_0) \to 0\} \tag{3.13}$$

The DA of a stable equilibrium in a nonlinear system is defined as the region of the state space. Each trajectory that originates from that region converges to equilibrium. It refers to the region where a given control law makes an equilibrium point asymptotically stable. Therefore, the system can be operated within the DA in a secure manner. The DA is also referred to in the literature as the region or the basin of attraction.

The stability and safety margin of systems concerning a specific beginning state can be easily predicted by estimating the DA. The correct planning of the nonlinear system's behaviour generally requires some understanding of its size and shape. The nonlinear system's transient stability is also enhanced by increasing the magnitude of DA. Lyapunov stability theory offers techniques for estimating the area of asymptotic stability with the goal of approximating the DA by a level set of a Lyapunov function around the equilibrium point.

A vector x_e presents a point at equilibrium for a dynamic system model $x(t) = f(t(\dot{x}(t))$ if once the state vector equals x_e and it continues equal to x_e for all time.
The equilibrium points satisfy

$$f(t, (\dot{x}(t))) = 0 \tag{3.14}$$

A point x_e is called an equilibrium point of $f(t,x)$ at time t_0 if for all $t > t_0$

$$f(t, x_e) = 0 \tag{3.15}$$

Due to the nonlinearity that CPL adds, equilibrium points may not exist when a large deviation occurs.

3.3.2 Robust Stability Analysis for DC microgrid connected to multiple CPL

The stability assessment of a DC microgrid is critical since the system equilibrium is dependent on unpredictable CPLs that change throughout the operating time in real-world applications, and an excessive CPL might induce instability. The methods for analysing the robust stability defined in 3.3.1 are developed in this subsection.

The technique uses Lyapunov theory to formulate the convex optimization problem and gives sufficient requirements for robust stability. This method is computationally efficient as it employs the LMI technique, which provides suitable settings for robust stabilisation of the nonlinear system. The approach addresses the problem and provides a reliable region for estimation.

The nonlinear system in the general form is stated as :

$$\dot{x} = Ax + f(t,x) \tag{3.16}$$

Here $x \in R^n$ denotes the system state, A is $n \times n$ constant state matrix and $f(.,.)$ signifies the nonlinearity in the equation. In the domains of continuity, the nonlinear function $f(.,.)$ given in equation(7) is assumed to be uncertain and satisfies the given quadratic inequality:

$$f^T(x)f(x) \leq \alpha^2 x^T H^T H x \tag{3.17}$$

Here $\alpha > 0$ is the bonding factor that determines the system's robustness. H denotes a constant matrix of $m \times n$ size. The large α value indicates additional robustness to the nonlinear disturbances. $f^T(x)$ represents the transpose of function matrix $f(x)$.

Additionally, the inequality in (3.17) defines the piecewise continuous functions, and for each given function, H is defined as:

$$H_\infty = \{f: R^{n+1} \to R^n | f^T f \leq \alpha^2 x^T H^T H x \ in \ the \ domains \ of \ continuity\} \tag{3.18}$$

The function H_∞ is known as a matrix-valued function for hardy space, which denotes the largest gain in every direction in state space. The objective is to create stability around the equilibrium. If the trajectories starting at the equilibrium point will approach the region of equilibrium, then the trajectories are believed to be stable asymptotically near equilibrium. Therefore, if the equilibrium provides global asymptotic stability for all $f(.,.) \in H_\infty$, the nonlinear system given in (1) will be stable robustly with degree α. The function $f(.,.)$ is dependent on the system's state x rather than its output. A Lyapunov stability theorem addresses the system's robust stability[100].

Let $V(x)$ be a continuously differentiable real-valued Lyapunov function specified in a region $R(0) \subseteq R^n$ and expressed as :

$$V(x) = x^T M x \tag{3.19}$$

Here M is a symmetric positive definite matrix. If all of a symmetric matrix's eigenvalues are positive, the matrix is said to be positive definite. The system stated by (3.16) is asymptotically stable for $\left(\lim_{t \to \infty} x(t) = 0, \forall x_0 \neq 0\right)$ if

- $V(x) = 0$ and
- $\dot{V}(x) < 0$ while the DA around the equilibrium is described as :

$$f(.,.) = \{x \in R^n : V(x) = \gamma , \gamma > 0\} \tag{3.20}$$

The stability theory can be established by proving the negative definiteness of the Lyapunov function's derivative in the domain of continuity of $f(.,.)$ given in(3.19).

$\dot{V}(x)$ is calculated using (3.19) to verify the system's stability and is expressed as:

$$\dot{V}(x) = x^T(A^T M + MA)x + f^T M x + x^T M f \tag{3.21}$$

Further, $\dot{V}(x)$ must be less than zero for the system to remain stable, which occurs only when

$$x^T(A^T + MA)x + f^T M x + x^T M f < 0 \tag{3.22}$$

Though this condition does not show convexity in M, an analogous condition may be produced by changing variables and using the S-procedure, which can then be expressed as an LMI. Thus, the following LMI can be used to explain the sufficient and necessary requirements for the presence of the quadratic Lyapunov function:

$$\begin{cases} M > 0 \\ \begin{bmatrix} AM + YM^T + \tau\alpha^2 H^T H & I \\ I & -I \end{bmatrix} < 0 \end{cases} \tag{3.23}$$

Where τ is the constant of s-procedure, which cannot be negative. If $\alpha^2 = \frac{1}{\gamma}$ and $M = \tau Y$, the equation (3.23) can be revised as follows by applying the compliment formula by Schur:

$$\begin{cases} Y > 0 \\ \begin{bmatrix} AY + YA^T & I & YH^T \\ I & -I & 0 \\ HY & 0 & -\gamma I \end{bmatrix} < 0 \end{cases} \tag{3.24}$$

The matrix is selected in a way that maximises the constraint in order to assure robust stability over a wide class of H_∞. This can be done by solving the LMI problem in Y, as illustrated below.

$$\left\{ \begin{array}{c} \textit{Minimize } \gamma \\ \textit{Subject to } Y > 0 \\ \begin{bmatrix} AY + YA^T & I & YH^T \\ I & -I & 0 \\ HY & 0 & -\gamma I \end{bmatrix} < 0 \end{array} \right. \tag{3.25}$$

The system (3.16) is shown to be robustly stable with degree α by the LMI condition's viability.

3.4 Control Design

3.4.1 Control Design using Hardy Space of matrix-valued function

Any control system essentially requires robustness. Also, a linear control system cannot be used in all situations. The controller, in this case, must be designed using nonlinear control approaches. The nonlinear H-infinity control theory can be used to create nonlinear robust control laws. The H-infinity nonlinear control theory [67] provides a technique to develop the approach for a robust nonlinear controller design. However, to utilise the Hamilton-Jacobi-Isaacs (HJI) inequality theory, a first-order nonlinear differential inequality ought to be resolved to know the implications of the concept. This inequality is the nonlinear counterpart of the well-known Riccati inequality.

In contrast to the Riccati inequality, the HJI inequality has no systematic procedures for solving it. In reality, using the assumptions of the H-infinity nonlinear control theory is constrained by this limitation. Most of the approaches to this problem described in the literature [68],[69] rely on applying complex numerical techniques to approximate the HJI equation. These are, however, simply rough estimates of the available options.

Furthermore, except [70], almost none of these methods provide quantitative information regarding the extent of the state space region where the calculation is valid. In contrast to earlier studies, this strategy approaches the nonlinear H-infinity control problem from an inequality perspective. This method makes it simple to solve the H-infinity nonlinear control problem. The suggested strategy also offers an estimate of the state space region where this solution is practicable in addition to a solution to the problem.

This method generates robust stability for a system and estimates the suitable stability region using a quadratic Lyapunov function $V(x)$. The challenge of finding an answer to the H-infinity nonlinear control technique can be viewed as a question of selecting a quadratic Lyapunov function as a positive scalar function given as:

$$V(x) > 0 \tag{3.26}$$

The design for the controller of a system is essential because it stabilizes the system during disturbances by providing feedback. In addition to stabilising the system, a good choice of feedback maximises the system's tolerance of any unknown nonlinear disturbance by raising the DA. Following the use of the feedback, the closed-loop system equation converts as:

$$\dot{x} = \hat{A}x + f(.,.) \tag{3.27}$$

Where the closed-loop matrix is designated as $\hat{A} = A + BK$ and the control input is stated as:

$$u(x) = Kx \tag{3.28}$$

Here K represents a matrix of $n \times m$ size for constant gains. Developing feedback aims to increase the uncertainty bound α and stabilise the system.

Feedback control can effectively stabilise the system (3.27) if the closed-loop system provides robust stability with degree α. The equation for closed-loop is derived by replacing the open-loop system matrix A in (3.27) with the closed-loop matrix $A + BK$, and is written as:

$$\left\{ \begin{array}{c} \textit{Minimise } \gamma \\ \textit{Subject to } Y > 0 \\ \begin{bmatrix} (A + BK)Y + Y(A + BK)^T & I & YH^T \\ I & -I & 0 \\ HY & 0 & -\gamma L \end{bmatrix} < 0 \end{array} \right. \tag{3.29}$$

After simplifying, it can be written as :

$$\left\{ \begin{array}{c} \textit{Minimise } \gamma \\ \textit{Subject to } Y > 0 \\ \begin{bmatrix} AY + YA^T + BKY + YK^T B^T & I & YH^T \\ I & -I & 0 \\ HY & 0 & -\gamma L \end{bmatrix} < 0 \end{array} \right. \tag{3.30}$$

The aforementioned equation might not, however, take the form of LMI in Y and K. It is possible to convert the equation into an LMI problem by changing the variable:

$$KY = L \text{ or } K = LY^{-1} \tag{3.31}$$

To make the system viable, the gain matrix size should be controlled to ensure the right value of $\bar{\alpha}$. The system is configured as follows:

$$\begin{cases} L^T L < K_l I, K_l > 0 \\ Y^{-1} < K_y I, K_y > 0 \end{cases} \tag{3.32}$$

Where K_y, K_l are the norm of bounding for the matrix of gain K.

The appropriate value of α can be ensured by taking $\gamma = \dfrac{1}{\alpha^2}$ and need that

$$\gamma - \frac{1}{\bar{\alpha}^2} < 0 \tag{3.33}$$

When all of the restrictions are added together, the ultimate optimization problem becomes:

$$\begin{cases} \quad Minimise\ \gamma + K_l + K_y \\ \qquad Subject\ to\ Y > 0 \\ \begin{bmatrix} AY + YA^T + BL + L^T B^T & B & YH^T \\ B^T & -I & 0 \\ HY & 0 & -\gamma L \end{bmatrix} < 0 \\ \begin{cases} \gamma - \frac{1}{\bar{\alpha}^2} < 0 \\ \begin{bmatrix} -K_l I & L^T \\ L & -I \end{bmatrix} < 0 \\ \begin{bmatrix} Y & I \\ I & K_y I \end{bmatrix} > 0 \end{cases} \end{cases} \tag{3.34}$$

If (3.33) is feasible, the design approach will be stabilizable robustly to a degree α via the proposed control rule. H∞ controller algorithm using LMIs can be written as:

Suppose the state matrices A, B, C, and D are available.

1: Find L, Y ∈ Sn that meet (3.31).

2: Calculate M

3: To obtain the system matrix K for a controller, solve (3.34).

*Sn = symmetric matrices' positive definite cone

3.4.2 DC microgrid framework with multiple CPL:

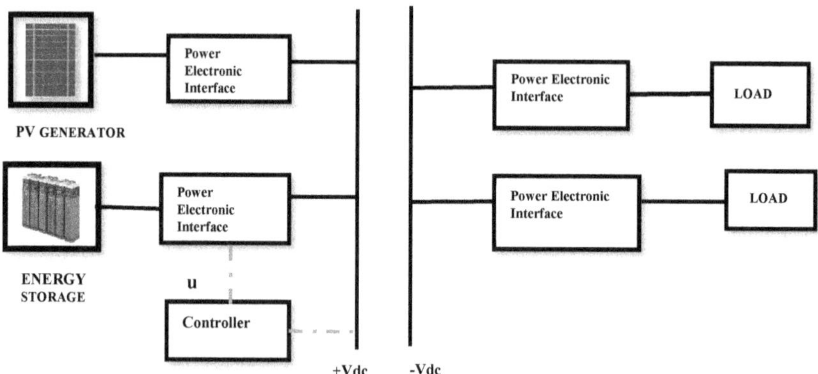

As displayed in Fig. 3.2, we investigate a control approach for a DC microgrid using several CPLs in this work. The suggested method is used to provide stability during disturbances and also to calculate the DA for multiple CPLs. Based on (3.7), the ith bus dynamic model in the system, displayed in Fig.3.1, can be stated as:

$$\dot{x}_i = A_i x_i + b_i f_i + A_{is} x_s \tag{3.35}$$

Where $i \in I_n \triangleq \{1,2,......n\}$ and $s \triangleq n+1$ represents the source. $x_s = [x_{1,s} \ x_{2,s}]^T = [i_{L,s} \ v_{C,s}]^T$ and $x_i = [x_{1,j} \ x_{2,j}]^T = [i_{L,i} \ v_{C,i}]^T$ are the source and CPL's state vectors, respectively.

The state matrices can be calculated as:

$$A_i = \begin{bmatrix} -\dfrac{r_i}{x_i} & \dfrac{-1}{L_i} \\ \dfrac{1}{C_i} & 0 \end{bmatrix}, b_i = \begin{bmatrix} 0 \\ \dfrac{1}{C_i} \end{bmatrix}, A_{is} = \begin{bmatrix} 0 & \dfrac{1}{L_i} \\ 0 & 0 \end{bmatrix} \tag{3.36}$$

Referring to (3.7), the source subsystem is presented as:

$$\dot{x}_s = A_s x_s + b_s V_s + b_{bat} i_{bat} + \sum_{i=1}^{n} A_{cn} x_i \tag{3.37}$$

Where

$$A_s = \begin{bmatrix} -\dfrac{r_s}{x_s} & \dfrac{-1}{L_s} \\ \dfrac{1}{C_s} & 0 \end{bmatrix}, b_s = \begin{bmatrix} \dfrac{1}{L_s} \\ 0 \end{bmatrix}, A_{is} = \begin{bmatrix} 0 & 0 \\ \dfrac{-1}{C_s} & 0 \end{bmatrix}, b_{bat} = \begin{bmatrix} 0 \\ \dfrac{-1}{C_s} \end{bmatrix}$$

The DC microgrid dynamic model is reformulated as follows, considering a coordinate shift close to equilibrium and taking the current i_{bat} from energy storage to provide the control input:

$$\dot{\tilde{x}} = A\tilde{x} + B_i f_i + B_{bat} \tilde{i}_{bat} + B_s V_s \tag{3.38}$$

Here, the state variable x and other system matrices are defined as:

$$x = (x1, x2,.....xn)^T, A = \begin{bmatrix} A_1 & . & . & . \\ . & A_2 & . & . \\ . & . & A_n & . \\ A_{cn} & . & A_{cn} & A_s \end{bmatrix}, B_i = \begin{bmatrix} b_1 & . & . & . \\ . & b_2 & . & . \\ . & . & . & . \\ . & . & . & b_n \\ 0 & 0 & 0 & 0 \end{bmatrix}, B_{bat} = \begin{bmatrix} 0 \\ 0 \\ b_{bat} \end{bmatrix},$$

$$B_s = \begin{bmatrix} 0 \\ 0 \\ b_s \end{bmatrix}$$

and the nonlinear function is given as $f_x = (f_1,.....f_n)^T$

Where

$$f_i(\tilde{x}_i) = \frac{\tilde{x}_i}{x_{io}(\tilde{x}_i + x_{io})} \ , \ i \in I_n$$

The state of the *ith* CPL is represented by x_i, and the operating position for that state is represented by x_{i0}. The nonlinear term $f_i(\tilde{x}_i)$ is considered to fulfill a quadratic bond stated as in (3.37). The goal is to use feedback to stabilise a DC microgrid with many CPLs while simultaneously raising the boundary parameter α.

The equation(3.33) is rewritten as :

$$\left\{ \begin{array}{c} Minimise\ \gamma + K_l + K_y \\ Subject\ to\ Y > 0 \\ \begin{bmatrix} AY + YA^T + B_{bat}L + L^TB^T & B_{bat} & YH^T \\ B^T & -I & 0 \\ HY & 0 & -\gamma L \end{bmatrix} < 0 \\ \left\{ \begin{array}{c} \gamma - \frac{1}{\bar{\alpha}^2} < 0 \\ \begin{bmatrix} -K_l I & L^T \\ L & -I \end{bmatrix} < 0 \\ \begin{bmatrix} Y & I \\ I & K_y I \end{bmatrix} > 0 \end{array} \right. \end{array} \right. \qquad (3.39)$$

Where the gain matrix $K = LY^{-1}$. The bounding norms K_l , K_y are set as: $K < \sqrt{K_l\,K_y}$. The flowchart for planning the control design is shown in Fig. 3.3.

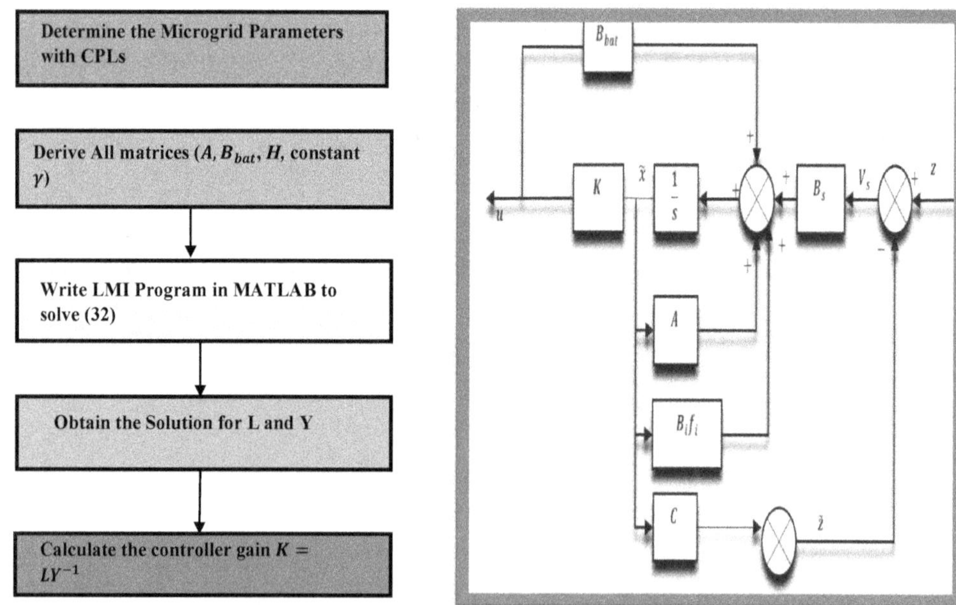

Fig.3.3: Flow chart for designning the proposed Controller   Fig.3.4: Block diagram of H_∞ Controller

The H_∞ controller design block diagram is displayed in Fig.3.4.

3.5 Simulation and findings :

This section aims to demonstrate the usefulness of the controller design suggested in section 3.4. The efficacy of the suggested method was analysed by simulations conducted on MATLAB/Simulink. The proposed design technique is simulated with many CPLs using the parameters provided in Table -3.1.

The case study is performed considering two scenarios: (i) a load-varying condition and (ii) an uncertain fault condition. Fig. 3.2 represents the simulation diagram. The PV system used for simulation has a rating of 10 KW together with a nominal operating temperature of 298K with 1000 W/m^2 of solar irradiance. The considered bus voltage is taken at 200 V. The ESS is attached in parallel to the PV system and has a nominal DC voltage of 48V with 150 Ah capacity.

TABLE -3.1

Simulation System Parameters for Multiple CPLs

R1 = 0.8Ω	R2 = 0.42Ω	Vdc = 200V
L1 = 40 mH	L2 = 19.5 mH	Rs = 0.4Ω
C1 = 1 mF	C2 = 1.05 mF	Ls = 17.3 mH
P1 = 900W	P2 = 600 W	Cs = 1.05 mF

Initially, the robustness index α_n is calculated using equation(3.29). The same robustness index is calculated using equation(3.34). The value of both the calculated robustness indices is compared in Table 3.2.

TABLE -3.2

Comparison of robustness index

Robustness index	Calculated value
α_n (S)	0.024
α(C)	0.099

The robustness index value indicates that the proposed controller design reduces the size of the DA while optimising the gain matrix value. The state variables were ensured to be within the stability domain and deliver a viable result. The DA is computed assuming $x \in R^n$. If there occurs a diagonal matrix $M \geq 0$ such that it fulfills the LMI equations, then the DA is assessed as :

$$\Omega(r) = \{x \in R^n \mid V(x) = x^T M x \leq r\} \tag{3.40}$$

The suggested algorithm for determining the DA has the advantage that the LMIs do not grow exponentially as the number of CPLs increases. Figure 3.5 shows a comparison of DAs with and without the proposed controller. It has been observed that increasing the robustness index increases the DA.

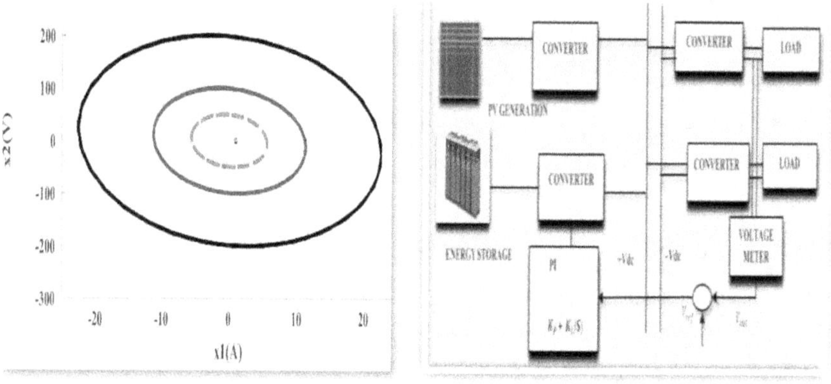

Fig. 3.5: Comparison of DA without and with the proposed controller (proposed approach is presented by the black line and DA without control is presented by the dashed red line)

Fig. 3.6: Block diagram of PI controller

Case I –load varying condition:

At t=0.5 seconds, the load of the PV system is increased while it is functioning at STC irradiance and temperature. The load was operating at 900 W, and another load of 600 W is linked to the grid. For comparison, a traditional PI controller is displayed in Fig.3.6 to aid in assessing the suggested controller design performance.

The suggested control method is then employed for simulation using the same conditions, and its performance is assessed with the conventional PI controller. Gain Kp and Ki for PI controller are assessed as given in [34].

Fig. 3.7-3.9 depict the controller's system responses. A considerable load change can create a big disturbance in the voltage output.

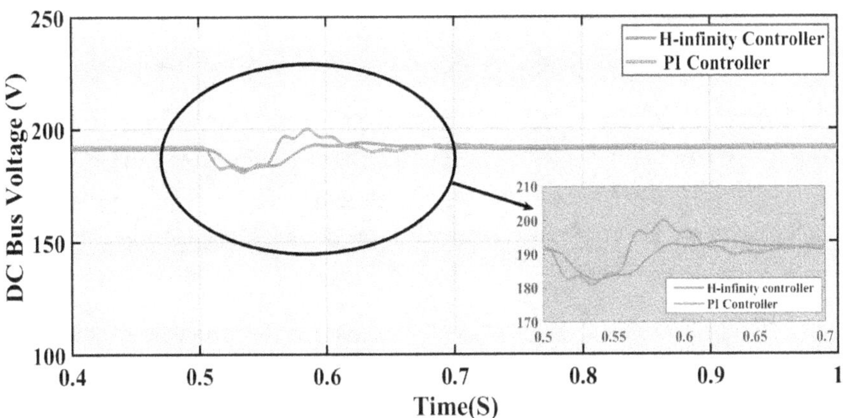

Fig. 3.7: DC Bus voltage during varying load demand

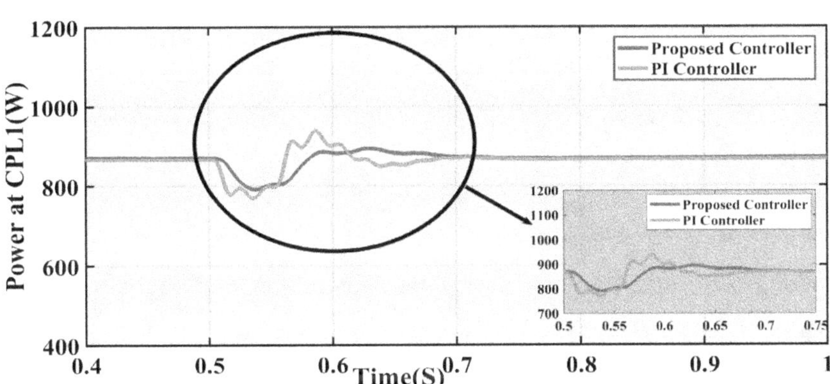

Fig. 3.8: Power at CPL1 under varying load demand

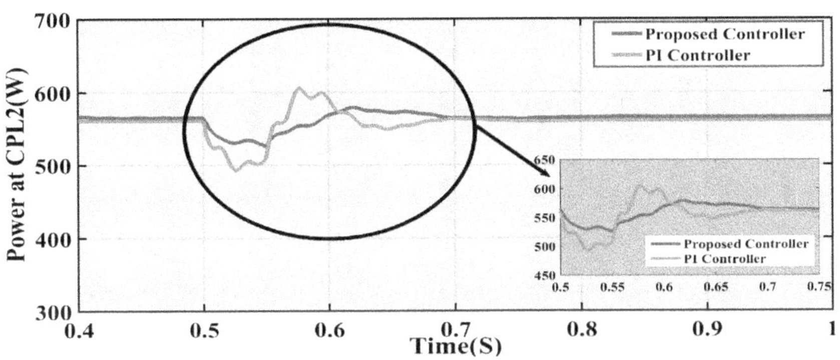

Fig. 3.9: Power at CPL2 under varying load demand

Case II – a Fault condition (Switching fault):

The system is operating at the specified values presented in Table -1. To assess the functioning of the suggested controller, a fault is introduced for 0.2s at the DC link. If the error is not addressed, the system will become unstable. Fig.3.10-3.13 shows the output voltage and current at CPL1 and CPL2, respectively. Bus voltage during fault is displayed in Fig.3.14. Load current and voltage changes are examined and compared for the suggested and conventional PI controller during fault conditions.

Fig. 3.12-3.13 exhibit both the controller's performances. The comparison demonstrates that the suggested controller can adjust the current very fast, maintaining a lesser overshoot than the traditional PI controller. At the time of disturbance, the proposed controller is operated by the ESS. Fig. 3.14 shows that even though both controllers can stabilize the voltage, the traditional PI control produces a significant transient response and converges over a long time compared to the suggested approach. The long settling time sometimes leads to system failure.

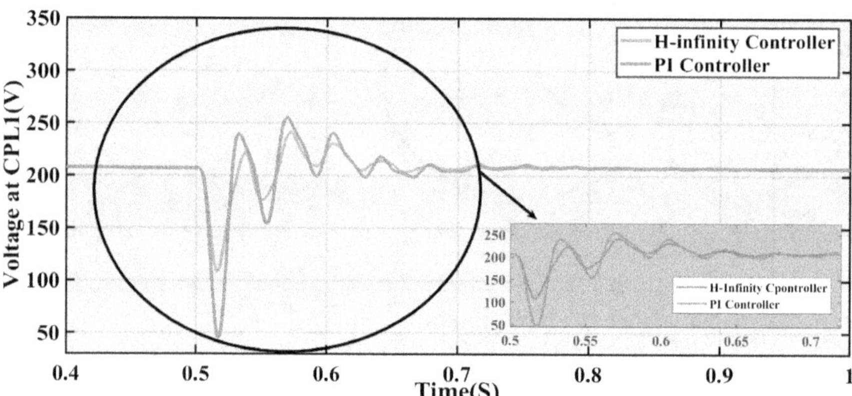

Fig.3.10: Voltage at CPL1(V) during a fault condition

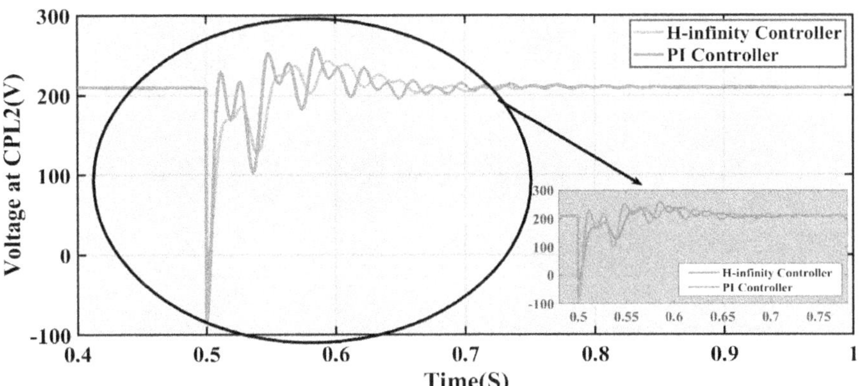

Fig.3.11: Voltage at CPL2(V) during a fault condition

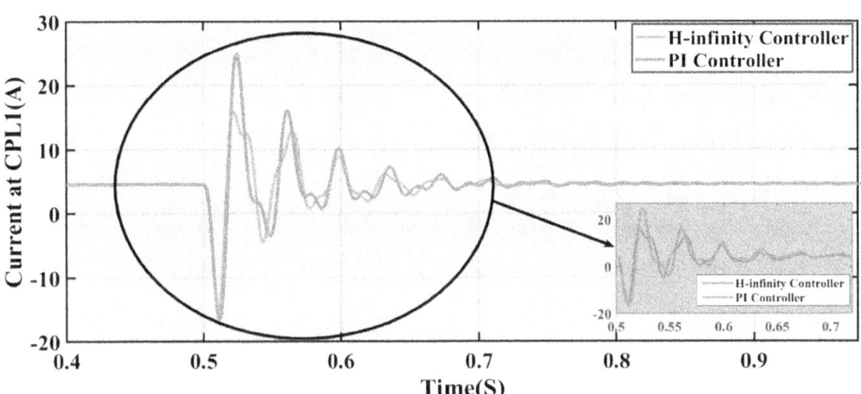

Fig.3.12: CPL1 current (fault condition)

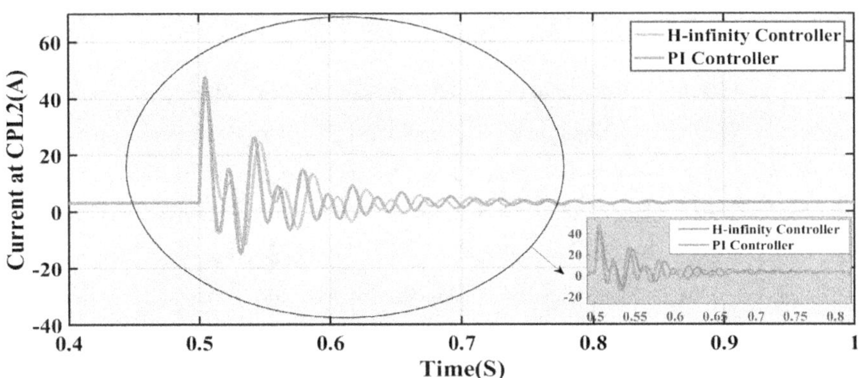

Fig.3.13: CPL2 current (fault condition)

44

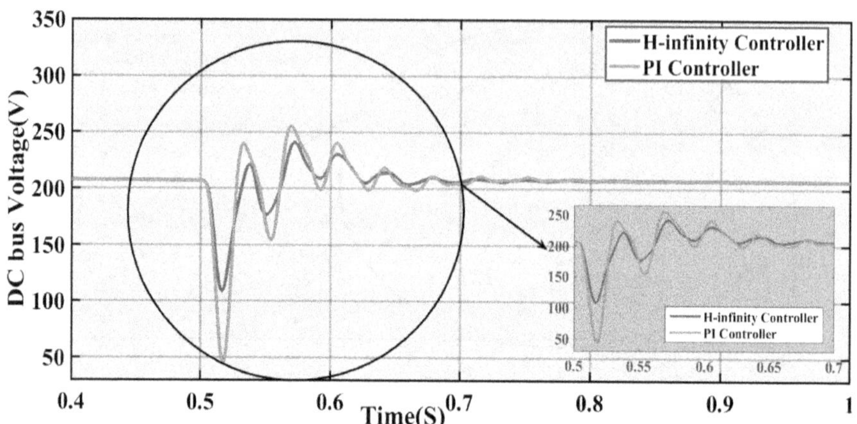

Fig.3.14: DC Bus Voltage during a fault condition

The system's responses show that the proposed approach aids the system in fast recovering from these perturbations. It is demonstrated that the recommended controller presents an improved performance than the PI, Sliding mode controller, and Fuzzy PID in terms of settling time, maximum overshoot, and steady-state errors.

TABLE – 3.3

Time Response of various methods for DC microgrid control

Controller Design methods	Settling time(s)	%overshoot	% Steady state error
PI	1.6091	0.326	0.9231
Fuzzy PID	1.3609	0.132	0.2325
Sliding Mode	0.5	0.46	-
Proposed design	0.14	0.245	0.021

Table 3.3 presents the comparison of the output of the proposed controller design using sliding mode controller, fuzzy PID controller, and traditional PI controller design approach. The settling times for PI, sliding mode, and fuzzy PID approach are 1.6091, 0.5, and 1.3609, respectively, while For the proposed technique, the corresponding observed value is 0.14. Comparisons are made between steady-state errors and percentage overshoots for the same. The suggested controller has a percentage steady-state error of 0.021, whereas the PI and Fuzzy PID controllers have estimates of 0.9231 and 0.2325, respectively. The comparison shows that the suggested approach has the quickest settling time and lowest steady-state error compared

to previous approaches. PI, Sliding mode, and Fuzzy PID, along with the suggested controller, have percentage overshoot values of 0.326, 0.46, 0.132, and 0.245, respectively. Figure 3.15 compares the transient responses of various techniques, including the proposed strategy.

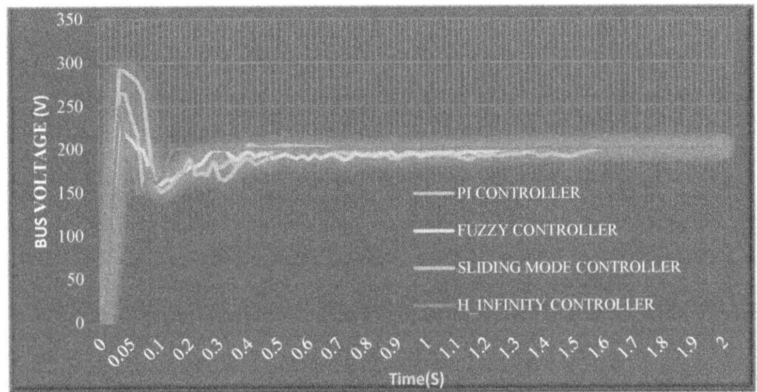

Fig.3.15: DC Bus Voltage Transient Response

In comparison to the PI and Sliding mode controllers, the suggested controller has less overshoot. As a result, it can be inferred that the proposed approach outperforms the sliding mode control, PI, and fuzzy PID approaches.

3.6 Conclusion:

This work proposes a novel approach to controller design created on norms of the H-infinity. The nonlinearity entailed by the CPL is considered in the model of the controller design. Also, the stability analysis is done using Lyapunov criteria. The Lyapunov criteria for stability hold are suitable for the nonlinear function. Simulations in MATLAB using the LMI solver were used to test the efficacy of the proposed approach.

The case study is carried out using the parameters listed in Table -3.1. The suggested controller's performance was assessed under two scenarios: (i)changing load conditions and (ii) fault conditions. Large fluctuations during a sudden variation in load or unintentional fault might cause voltage instability in the system. Under these conditions, the suggested approach stabilises the system quickly and enhances transient responsiveness.

The findings of the suggested controller are compared to the PI, sliding mode, and Fuzzy PID controllers for steady-state error, settling time, and percentage overshoot. The steady-state and transient response is significantly enhanced over existing methods. Since the time derivatives

46

of states are not required in the suggested method, the technique is robust against the disturbances compared to other control approaches. This strategy is more resistant to uncertainties than other linear approaches because it uses a nonlinear function.

Furthermore, by raising the DA, the proposed technique provides global stabilisation. The formula shows how to apply the estimation method for calculating gain to provide feedback to get the required robustness index with the maximum norm of gain value. By increasing the robustness index, the area of the DA can be increased to the appropriate amount, improving system control. The size of the DA or region for stability will improve as the robustness index increases. According to the simulation findings, the suggested controller with CPLs has greater closed-loop performance than a conventional PI controller. The closed-loop method for DC microgrids with CPLs has substantially improved its DA.

The suggested controller minimises the requirement of the injecting current while applying it to the DC microgrid via energy storage, thus increasing the system's performance. The suggested approach performs well during dc bus voltage faults and disturbances and addresses the challenges during load changes in the DC microgrid. The key benefit of the proposed strategy is that it reduces complexity while increasing the number of CPLs. As a result, working a DC microgrid with numerous CPLs can be improved with less complexity.

CHAPTER 4

ADAPTIVE CONTROL FRAMEWORK:

4. Adaptive Framework for CPL-connected DC Microgrid :

The high level of interaction between distinctly distributed generators, each with its interface and controller characteristics, can have an adverse impact on the microgrid's global stability. Because of the widespread use of converter-interfaced DERs, most microgrid stability issues are caused by the dynamics and control of their converters. Many power electronic inverters and motor drives act like CPL when tightly controlled. CPL exhibits negative incremental impedance characteristics, as shown in Fig.1.3, offers a negative relationship between voltage and current variations, resulting in negative incremental impedance instability. Also, the control laws in a microgrid may experience frequent topology changes and load switchings; hence, it should be responsive to those uncertainties.

Control design and analysis for systems with static and dynamic uncertainties and systems operating in the presence of environmental disturbances and other unknown elements are difficult. When a perfect model of a physical system is unavailable due to parametric uncertainties, an adaptive technique can be used to detect the unknown constraints. One primary aspect distinguishing adaptive control is that it analyses the overall plant's performance and uses the data to automatically update the control rule to improve performance. In most circumstances, a nonlinear physical system's exact model is unavailable.

An adaptive design can be developed for unknown factors in the dynamic system. The adaptive control approach satisfies this application by reducing uncertainties using parameter estimation. If the system parameters change later, adaptive control adjusts to account for these new conditions.

The new trend in nonlinear system control is to employ intelligent adaptive control techniques such as fuzzy logic techniques, neural networks, artificial intelligence algorithms, machine learning, etc. ANFIS is a popular intelligent control system that combines the features of fuzzy logic (FL) and neural networks (NN) and was built in an adaptive network architecture. It is frequently used for complex and nonlinear systems in many sectors. As the DC microgrid operates with non-linear uncertain loads, the ANFIS' capacity to handle uncertainty has proven beneficial.

This chapter presents work on an ANFIS-based nonlinear adaptive controller to stabilise a DC microgrid linked to dynamic CPLs. In real-time operation, the microgrid feeds uncertain and unpredictable CPLs. Hence, There is a need to measure the instantaneous power value for uncertain dynamic CPLs to enhance the system stability. Measuring all data from all units in large DC microgrids necessitates a large number of measurement equipment, which is not cost-effective. As a result, state estimation methods for predicting uncertain data are preferable for improving DC microgrid stability[71].

There are two basic approaches for estimating the uncertain parameter:

(i)   Stochastic approach

(ii)   Deterministic approach

The computing burden of stochastic approaches is significant, making them incompatible for practical purposes.

Cubature Kalman filters (CKF) are a type of deterministic sampling approach with several advantages, including accuracy, reduced processing effort, and increased numerical stability[72]. Therefore, the cubature Kalman filtering is a likely option for the prevailing uncertainty on DC microgrids and faulty measurements in power systems and power electronics. In this chapter, a CKF is used to solve an estimation challenge to evaluate the states of a DC microgrid and the power of a dynamic CPL. The predicted value is afterward used in an ANFIS controller to control the ESS current effectively. For a large range of variations in power, the suggested approach effectively controls the transient response of the DC microgrid.

4.1 Dynamic Model of DC Microgrid with uncertain CPL Power:

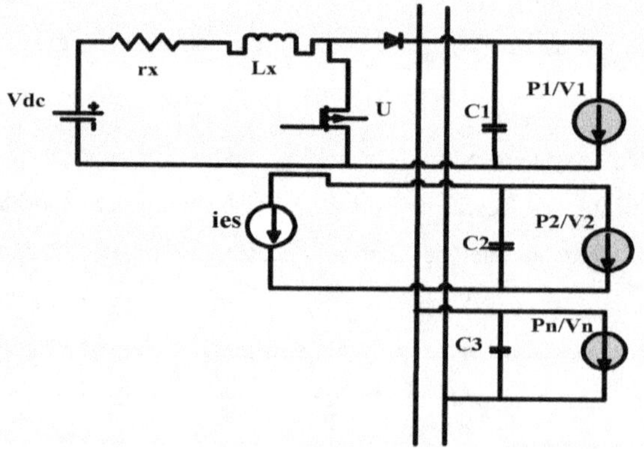

Fig.4.1: Simple DC microgrid circuit schematic with n CPLs

Fig.4.1 displays a simplified DC microgrid circuit diagram connected to multiple CPLs. The diagram shows multiple CPLs are connected to a solar panel and an ESS. The overall state-space equation of the DC microgrid displayed in Fig.4.1 is presented as:

$$\dot{X} = \bar{A}X + \bar{D}P + B_s V_{dc} + B_{es} i_{es} \tag{4.1}$$

Where $X = [x_1^T \; x_2^T \ldots \ldots x_i^T \; x_s^T]^T$ are the state vectors and $P = [P_1 \; P_2 \ldots \ldots P_n]^T$ are the unknown CPLs. The system matrices are expressed as:

$$\bar{A} = \begin{bmatrix} A_1 & \cdots & 0 & A_{1j} \\ 0 & A_2 & 0 & A_{2j} \\ \vdots & \ddots & 0 & \vdots \\ 0 & 0 & 0 & A_{ij} \\ A_{cn} & \cdots & A_{cn} & A_s \end{bmatrix}, \; \bar{D} = \begin{bmatrix} d_1 & \cdots & 0 \\ \vdots & \ddots & \vdots \\ 0 & \cdots & d_i \\ 0 & 0 & 0 \end{bmatrix}, \; B_s = \begin{bmatrix} 0 \\ \vdots \\ 0 \\ b_s \end{bmatrix}, \; B_{es} = \begin{bmatrix} 0 \\ 0 \\ \vdots \\ b_{es} \end{bmatrix}$$

V_{dc}, i_{es} are source voltage and ESS current, respectively. B_s, B_{es} are input matrices for source and ESS, respectively, and D represents the output matrices for CPLs. The goal is to create a controller that guarantees closed-loop performance, including DC microgrid stability, by estimating the CPL power with uncertainties.

4.1.1 CPL Power Model With Uncertainty :

Almost all control system design methodologies revolve around mathematical modelling. No physical system, however, can be completely modelled, and there will always be some misalignment between the design and the physical structure. Even though an exact structure of the physical system is provided, some exogenous disturbances can create challenges in the control system design through inadequate measurement of either state or output and cause uncertainty [73].

Other forms of uncertainty may exist as a result of a lack of understanding of the physical system. Even with a perfect model, obtaining precise parameter values for the physical system is difficult. The state-space model of nonlinear systems with uncertain parameters can be presented as :

$$\begin{cases} \dot{x} = f(x,\theta) + g(x,\theta)u \\ \quad\;\; y = h(x) \end{cases} \tag{4.2}$$

Where $x \in R^n$ is the state vector, $\theta \in S$ is the unknown vector with $S \subset R^p$ being a compact set, $f, g : W_x \subset R^{n \times p} \to R^n$ are suitably smooth vector fields, $h: W \subset R^n \to R$ denotes a scalar function, $u \in R$ denotes the input, and $y \in R$ denotes the output. W presents the DA around the equilibrium. The vector fields f, g are presented as :

$$\begin{cases} f(x,\theta) = \sum_{k=1}^{p} \theta_k f_k(x) \\ g(x,\theta) = \sum_{k=1}^{p} \theta_k g_k(x) \end{cases} \tag{4.3}$$

Here, $f_k, g_k : W_x \subset R^n \to R^n$.

Using an approach that estimates the uncertain parameter and leads to convergence as time passes is necessary.

4.1.2   State Space Model for pth bus DC Microgrid

The DC microgrid's state-space equation with multiple CPLs is written as:

$$\begin{cases} \dot{x}_p = A_p x_p + d_p P_p + A_{ps} x_s \\ \qquad \dot{y}_p = c_p x_i \end{cases} \tag{4.4}$$

Where $x_p = [i_{l,p}, v_{c,p}]^T$ is a state vector and represents the current and voltage for the system inductor and capacitor, respectively, for $p \in I_n \triangleq \{1,2, \ldots \ldots n\}$. Here n is the number of CPLs. Also, $x_s = [x_{1,s}, x_{2,s}]^T = [i_{L,s}, v_{c,s}]^T$ are the state vectors for the source. P_i is the pth CPL power, while y_p denotes the (n-dimension) measurement of the dynamical system. The matrices framed above can be explained as:

$$A_p = \begin{bmatrix} -\dfrac{r_{l,p}}{L_p} & \dfrac{-1}{L_p} \\ \dfrac{1}{C_p} & 0 \end{bmatrix}, d_p = \begin{bmatrix} 0 \\ -\dfrac{1}{C_p v_{c,p}} \end{bmatrix}, A_{ps} = \begin{bmatrix} 0 & \dfrac{1}{L_p} \\ 0 & 0 \end{bmatrix} \quad c_p = \begin{bmatrix} 0 & 1 \end{bmatrix}$$

The source subsystem with n number of CPL can be presented as

$$\begin{cases} \dot{x}_s = A_s x_s + b_s v_{dc} + b_{es} i_{es} + \sum_{p=1}^{n} A_{cn} x_p \\ \qquad y_s = c_s x_s \end{cases} \tag{4.5}$$

Where

$$A_s = \begin{bmatrix} -\dfrac{r_s}{L_s} & \dfrac{-1}{L_s} \\ \dfrac{1}{C_s} & 0 \end{bmatrix}, b_s = \begin{bmatrix} \dfrac{1}{L_s} \\ 0 \end{bmatrix}, A_{cn} = \begin{bmatrix} 0 & 0 \\ -\dfrac{1}{C_s} & 0 \end{bmatrix}, b_{es} = \begin{bmatrix} 0 \\ -\dfrac{1}{C_s} \end{bmatrix}, c_s = \begin{bmatrix} 0 & 1 \end{bmatrix}$$

The CPLs displayed in Fig.4.1 are connected to the DC bus using RLC filters, where CPLs are shown as a current sink [74]. The state-space equation of the pth state vectors is presented as [75]:

$$\begin{cases} \dot{i}_{L,p} = -\dfrac{r_i}{L_i}\, i_{L,p} - \dfrac{1}{L_p}\, v_{c,p} + \dfrac{1}{L_p}\, v_{dc} \\[2mm] \dot{v}_{c,p} = \dfrac{1}{C_p}\, i_{L,p} - \dfrac{1}{C_p}\dfrac{P_p}{v_{c,p}} \end{cases} \qquad (4.6)$$

4.1.3 Power estimation using the CKF approach

This section explains the CKF approach's design technique for estimating the CPL's unknown power. Although it is considered that an ideal CPL consumes constant power independent of the source voltage, practically, CPLs connected to a microgrid can be uncertain and unpredictable. Estimating the uncertain dynamic CPLs' instantaneous power is necessary to increase the stability performance of any control approach. This uncertain power can be estimated with the help of current and voltage sensing devices. The current sensor connected in series through CPLs reduces the ripple filtering effect and raises the impedance output.

Besides, addition of extra sensors increases system complexity, as well as the expenses of procuring, installing, and maintaining them. Therefore, the estimation methods for estimating the state vectors are better options than the sensor installation. Various estimation methods include NDO, Kalman filter, EKF, and CKF techniques.

The CKF approach has better convergence properties and filter stability. Also, it provides the solutions for the error covariance matrix. Importantly, CKF makes use of a non-product directive for sampling, whereas other approaches use a product law. These features encourage the use of the CKF in real-world applications.

This section explains the use of CKF for the estimation of states and CPL.

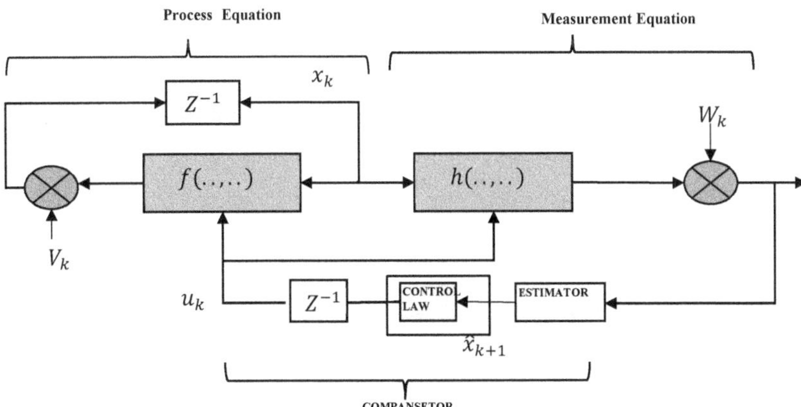

Fig. 4.2: An estimator-based feedback control is shown in the signal-flow diagram of a dynamic state-space model

Filtering means deriving the needed signal from a noisy obtained signal following the optimum error measures[76]. This chapter focuses on the cubature Kalman filter approach for the states and unknown power evaluation for the DC microgrid. This covers problems with state estimation posed by nonlinear dynamic systems with additive Gaussian noise. The physical state of a dynamic system explained using the term "state" in this context. The following equation represents the state evolution of a nonlinear system:

$$(state)_p = (state\ non\text{-}linear\ function)_{p-1} + (process\ noise)_{p-1}. \tag{4.7}$$

With p being the time index designating the value at p, the noise factor is intended to account for random disturbances brought on by power variations. As the measured data contains noise, another equation depicting the changing condition of the measurements can be expressed as follows:

$$(measurement)_p = ((non)linear\ function\ of\ state)_p + (measurement\ noise)_p. \tag{4.8}$$

Hence, filtering is described as the challenge of estimating the present state of the microgrid in the best possible manner when new states arrive in sequence with time. The following equations define the state-space equation of a nonlinear dynamic structure with disturbances:

$$\begin{cases} Process\ Equation: x_p = f(x_{p-1}, u_{p-1}) + v_{p-1} \\ Measurement\ Equation: z_p = h(x_p, u_p) + w_p \end{cases} \tag{4.9}$$

Here $x_p \in R^n$ is the dynamic system state, $u_p \in R^n$ introduces the identified control input derived from a compensator, as shown in Fig. 3.2, $z_p \in R^m$ is the output of the system. w_p and v_{p-1} are the measurement and process of Gaussian noise. R_p and Q_{p-1} presents the covariances and zero means, respectively. $f: R^{n_x}$ is the state functions for a system, $h: R^{n_z}$ is the measurement function of the system.

The posterior density of the state provides a complete statistical description of the state at that moment. The state's old posterior density can be updated as soon as a new measurement is received at that moment using two basic measures.

Time update: The predictive density is computed during the time update. The update equation is expressed as:

$$p(x_i|D_{i-1}) = \int p(x_i, x_{i-1}|D_{i-1})dx_{i-1} = \int p(x_{i-1}|D_{i-1})p(x_{i-1}|x_{i-1}, u_{i-1})dx_{i-1} \tag{4.10}$$

Where the history of input-measurement pairs is denoted by $D_{i-1} = \{ u_i, z_i \}_{i=1}^{k-1}$ up to time $i - 1$. $p(x_{i-1}|D_{i-1})$ denotes the earlier posterior density at time $i - 1$. $p(x_{i-1}|x_{i-1}, u_{i-1})$ provides the state transition density at the time $i - 1$. Measurement update: This comprises calculating the current state's posterior density.

$$p(x_i|D_i) = p(x_i|D_{i-1}, u_i, z_i) \tag{4.11}$$

A CKF is calculated by executing repetitive time, and measurement renews. The assessed state's final element is the projected CPL power. The CKF algorithm is divided into the steps below[77] :

- Beginning

The filter begins with the state initialisation and the covariance matrix:

$\hat{x} = E[x_0]$

$S_0 = chol\{[(x_0 - \hat{x}_0)\ (x_0 - \hat{x}_0)^T]\}$

where $chol\{\cdot\}$ explains the Cholesky factorization

Create Cubature point $\emptyset_i$

Create weight w_i

- **Time Update**

❖ Considering the original covariance and primary mean as $P_{j|j}$ and $\tilde{x}_{j|j}$ respectively are detected at the current time j, initially, factorize $P_{j-1|j-1}$ Cholesky breakdown as :

$$P_{j-1|j-1} = S_{j-1|j-1}S^T_{j-1|j-1} \tag{4.12}$$

Assess the cubature points for $(n = 1,2, \ldots \ldots m)$

$$X_{n,j-1|j-1} = S_{j-1|j-1}\tau_i + \tilde{x}_{j-1|j-1} \tag{4.13}$$

Here $m = 2k$

❖ Generate the calculated cubature points using a nonlinear model

$$X^*_{n,j|j-1} = F_x(X_{n,j-1|j-1}, u_{j-1}) \tag{4.14}$$

Estimate the predicted state

$$\hat{x}_{j|j-1} = \frac{1}{m}\sum_{n=1}^{m} X^*_{n,j|j-1} \tag{4.15}$$

❖ Calculate the expected error covariance

$$P_{j|j-1} = \frac{1}{m}\sum_{n=1}^{m} X^*_{n,j|j-1} X^{*T}_{n,j|j-1} - \hat{x}_{j|j-1}\hat{x}^T_{j|j-1} + Q_{j-1} \tag{4.16}$$

- **Measurement Update**
❖ $P_{j|j-1}$ is factorized by applying Cholesky decomposition

$$P_{j|j-1} = S_{j|j-1}|S^T{}_{j|j-1} \tag{4.17}$$

Compute the cubature points as :

❖ $X_{n,j|j-1} = S_{j|j-1}\tau_i + X_{j|j-1}$ for $n = 1, \ldots.,2n$ $\tag{4.18}$

❖ The calculated cubature points are transmitted through the measurement function:

$$Z_{n,j|j-1} = h(X_{n,j|j-1}, uj) \tag{4.19}$$

❖ The assessed measurement is expressed as:

$$\hat{Z}_{j|j-1} = 12k\sum Z_{nj|j-1} \tag{4.20}$$

❖ Estimate the auto-covariance matrix :

$$P_{zz,j|j-1} = \frac{1}{m}\sum_{n=1}^{m}(Z_{n,j|j-1} - \hat{Z}_{i,j|j-1}).\left(Z_{n,j|j-1} - \hat{Z}_{n,j|j-1}\right)^T + R_j \tag{4.21}$$

❖ Estimate the cross-covariance matrix :

$$P_{xz,j|j-1} = \frac{1}{m}\sum_{n=1}^{m}(X_{n,j|j-1} - \hat{X}_{n,j|j-1}).\left(Z_{n,j|j-1} - \hat{Z}_{n,j|j-1}\right)^T + R_j \tag{4.22}$$

❖ The Kalman gain is evaluated as:

$$K_j = P_{xz,j|j-1}P^{-1}{}_{zz,j|j-1} \tag{4.23}$$

❖ The vector of state is revised by applying the following law :

$$x_{j|j} = \hat{x}_{j|j-1} + K_j(z_j - \hat{z}_{j|j-1}) \tag{4.24}$$

❖ Calculate the corresponding Error Covariance as :

$$P_{j|j} = P_{j|j-1} - K_j P_{zz,j|j-1}K_j{}^T \tag{4.25}$$

The algorithm described above can be used to estimate DC microgrid states for a wide range of CPLs connected to it in real-time. The CPL's unknown power, *Pload*, is increased in the device states to achieve this. Where,

$$\dot{x} = [\dot{\iota L} \quad \dot{vC} \quad \dot{Pload}] = f(x,u) \tag{4.26}$$

The system calculation is given as :

$$y = \begin{bmatrix} I & 0 \\ 0 & 0 \end{bmatrix}\begin{bmatrix} iL \\ vC \\ Pload \end{bmatrix} = Hx \tag{4.27}$$

The state equations can be stated by taking the noise as w and v, for the system and measurement respectively, and adding up (4.26) and (4.27) together.

$$\begin{cases} \dot{x} = f(x,u) + w \\ \quad = Hx + v \end{cases} \tag{4.28}$$

The augmented state vector takes the resulting form:

$$\dot{z} = \begin{bmatrix} \dot{x} \\ \dot{P} \end{bmatrix} \tag{4.29}$$

Here x denotes the system's state vector, which includes inductor current and capacitor voltage, and P is the unknown power load. The CKF method, similar to previous Kalman filters, is iterative and must be executed at all repetition.

The suggested CKF requires a shorter computation time when compared to existing Kalman filters, such as unscented Kalman filters and particle Kalman filters. Additionally, matched to the unscented transformation and several other linearized techniques, the CKF outperforms the EKF and UKF in terms of performance and convergence, as the rules in CKF are quite precise. As a result, the CKF is employed to measure the DC microgrid states. The unknown power $Pload$ is obtained initiating the expected state vector $\hat{x}$ with CKF by using the state vector i_L, v_C.

4.2 Control Design using ANFIS-based Adaptive framework

As the name indicates, an adaptive network is presented in the network form of nodes and directional links that link the nodes together. The learning rule identifies the condition for these parameter changes to minimise a predefined error measure and the adaptive nodes, meaning that each output depends on the parameters connected with that node.

The fundamental principles governing learning in adaptive networks are gradient descent and chain rules. Fuzzy logic control (FLC) is an excellent technique for dealing with complex, non-linear, and ill-defined systems. Learning, adaptation, robustness, and speed are all-powerful capabilities of artificial neural networks (ANNs). The benefits of both the FLC and the ANN have been integrated into ANFIS. Regarding functionality, ANFIS is a form of adaptive network that functions similarly to a fuzzy inference system. The nonlinearity and parameter changes in the DC microgrid are solved using this control technology.

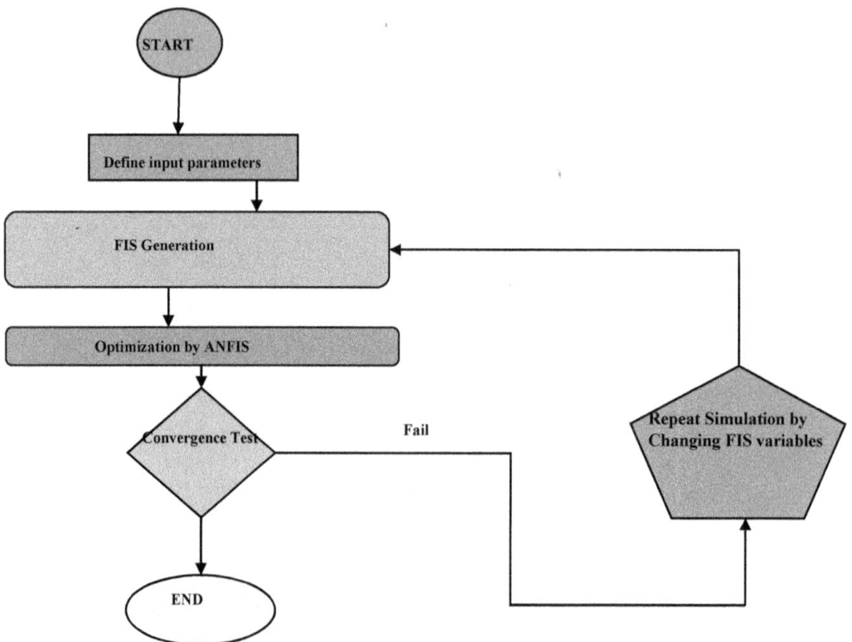

Fig. 4.3: Flow diagram showing the implementation Procedure of the ANFIS model

4.2.1   ANFIS Architecture for the Controller design

Artificial neural networks (ANN) and fuzzy inference systems (FIS) combined to create neuro-fuzzy techniques have gained popularity as a framework for tackling real-world problems. A fuzzy system that is made utilising a neural network learning technique serves as the foundation for a neuro-fuzzy system. This architecture can potentially capture the benefits of both the neural network and the fuzzy logic. FIS provides learning capability, while ANN provides the formation of a linguistic rule base.

Intelligent control has proven to be a viable alternative to model-based control techniques. This is because fuzzy logic and neural networks can more effectively deal with problems like ambiguity or unidentified plant characteristics and structure variations, increasing the control system's adaptability. ANFIS is made up of five layers and features a Multi-Input Single Output (MISO) structure. A basic ANFIS structure is shown in Fig.4.4, which has five layers: an input layer, three hidden layers, and one output layer. Directional links are used to connect the levels. x and y are two inputs, and f represents the single output. $A1$ and $A2$ represent the Membership Functions(MF) of input x, and $B1$ and $B2$ represent the MFs of input y.

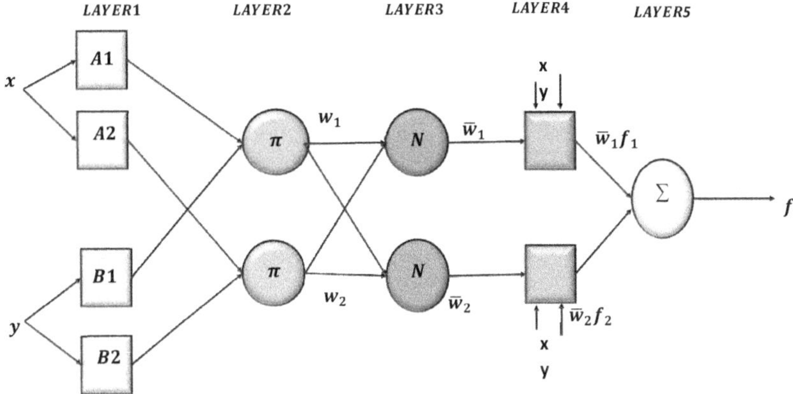

Fig.4.4: ANFIS Architecture

The AND operation is conducted with the Layer 1 outputs in the second layer, and the following rules are formed:

- *Rule 1: If x is A1 and y is B1, then $f1 = p1x + q1y + r1$,*
- *Rule 2: if x is A2 and y is B2, then $f2 = p2x + q2y + r2$*

Here p, q, r represents the output function parameters.

$p1, p2$: represents the number of membership functions used to fuzzify each input variable

$q1 \text{ and } q2$: represents the number of membership functions used to fuzzify the output variable.

$r1 \text{ and } r2$: represents the number of rules used in the rule base

Each output node is referred to as the firing strength of a rule. The firing strengths are normalised in the third layer.

A first-order Sugeno function with adaptive nodes is the output of the fourth layer. These nodes decide on the final parameters. The defuzzification layer calculates the crisp output of ANFIS and is the final layer. The output of the nodes in each tier of the five-layer ANFIS design is represented by o_i^k.

Layer1:

The fuzzification layer as first layer accepts an input and establishes acceptable member functions for it.

All node in the graph is denoted as a square node in the first layer, with the following node outputs:

$$o_i^1 = \mu A_i(x) = \exp\left(-\left(x - \frac{c_i}{\sigma_i}\right)^2\right) \tag{4.30}$$

Here, The membership function is presented as o_i^1, the node number is i, and the linguistic level is A_i. The membership function's centre and width are defined by c_i and σ_i respectively and are classified as presumed constraints.

Layer2:

Layer 1 signals are amplified by nodes in this layer. Nodes in Layer 2 are depicted as circular nodes. The output of Layer 2 can be written as:

$$o_i^2 = W_i = \mu A_i(x)\mu B_i(y) \tag{4.31}$$

In this case, W_i specifies the firing strength for all nodes.

Layer3:

Layer 3 features circular nodes and the number of layers and fuzzy rules are both the same. The following is the outcome:

$$o_i^3 = \widetilde{W}_i = \frac{W_i}{W_1 + W_2} \tag{4.32}$$

Layer4:

This layer's node i is an adaptive node with the following node function:

$$o_i^4 = \widetilde{W}_i f_i = \widetilde{W}_i(p_i x + q_i y + r_i) \tag{4.33}$$

Here, $\widetilde{W}_i$ is the output for the layer and p_i, q_i, r_i are succeeding parameters. These are known as subsequent parameters also.

Layer5:

The final output is calculated as the sum of all incoming signals by the sum marked fixed node:

$$o_i^5 = \sum \widetilde{W}_i f_i = \frac{\sum_i W_i f_i}{\sum_i W_i} \tag{4.34}$$

Basic Learning Rule for ANFIS:

The ANFIS controller must use a learning rule to set the premise and resulting parameters. The forward and reverse passes can be applied to set the parameters' values accordingly. The forward pass evaluates subsequent parameters while keeping premise parameters constant, whereas the reverse pass evaluates premise parameters while keeping subsequent parameters constant.

A hybrid algorithm can be used to train the data:

- Assume an adaptive network has L layers, with the kth layer having n nodes.
- The node in the ith point of the kth layer is denoted by (k, i)
- o_i^k represents the node task.

- Because the node's output is determined by its incoming signals and parameter sets (a, b, c), we have:

- $o_i^k = (o_i^{k-1}...., o_{n-1}^{k-1}, a, b, c)$

In this case, o_i^k is used as both a node task and a node output. Considering that the training data set has N input data, the error function for the nth $(1 \leq n \leq N)$ training data set is stated as follows:

$$En = \sum_{p=1}^{N}(Tp, n - O_{p,n}^N)2 \tag{4.35}$$

The sum of the squared errors is the error measure for the nth element. $O_p^n, T_p^n,$ and E_n are the pth, output, reference, and error components of the nth output vector, respectively. Consequently, the total measured error is as follows:

$$E = \sum_{n=1}^{N} E_n \tag{4.36}$$

The error rate at node (k, i) from (10) can be expressed as follows:

$$\frac{\delta E_n}{\delta o_{i,n}^k} = -2(T_{i,n} - O_{i,n}^k) \tag{4.37}$$

If σ is an ANFIS parameter, then the rate of change of error relating to σ is :

$$\begin{cases} \frac{\delta E_n}{\delta \sigma} = \sum_{O \epsilon S} \frac{\delta E_n}{\delta O} \frac{\delta O}{\delta \sigma} \\ \frac{\delta E}{\delta \sigma} = \sum_{n=1}^{N} \frac{\delta E_n}{\delta \sigma} \end{cases} \tag{4.38}$$

S signifies the nodes whose output is determined by σ. Additionally, $\frac{\delta E}{\delta \sigma}$ signifies the rate of change of the total measured error with σ. As a result, the revised generic parameter σ can be written as follows:

$$\Delta\sigma = -\beta \frac{\delta E}{\delta \sigma} \tag{4.39}$$

$$\beta = -\frac{J}{\sqrt{\sum_\sigma (\frac{\delta E}{\delta \sigma})^2}} \tag{4.40}$$

where β is the learning rate, and J is the step size, which can be changed to accelerate the convergence.

4.2.2 Design Approach for the Controller

The suggested observer-based adaptive controller, represented in Fig.4.5, is described in this segment. The suggested method allows for the creation of a controller and observer that ensures the desired disturbance rejection performance. The equations for the DC microgrid can be expressed in a canonical form as:

$$\dot{x}(t) = Ax(t) + Bu(t) + Bwd(t) \tag{4.41}$$

$$y(t) = Cx(t) \tag{4.42}$$

$$z(t) = C_z x(t) \tag{4.43}$$

The state vector, control input, and input disturbance of the system are denoted by $x(t), u(t)$, and $d(t)$. $A \in R^{nxn}$ for state matrix, $B \in R^{nxm}$ for input matrix, $Bw \in R^{nxq}$ for the input disturbance matrix, $C \in R^{mxn}$ for output matrix, and $Cz \in R^{mxn}$ which is equivalent to matrix C, are the system constant matrices presented in (4.41)-(4.43). An adaptive ANFIS-based State-feedback controller is applied to meet the control objectives described in section 3.2.1 with guaranteed performance. However, adopting a state-feedback system necessitates the measurement of all states in the system (4.41). As a result, an adaptive observer-based controller (4.41)-(4.42) with a state-feedback control rule (4.43) is utilized to lessen the necessary number of measurements and related installed sensors.

$$x(t) = A\hat{x}(t) + Bu(t) + L\left[y(t) - \hat{y}(t)\right] \tag{4.44}$$

$$\hat{y}(t) = C\,\hat{x}(t) \tag{4.45}$$

$$u(t) = -K\,\hat{x}(t) \tag{4.46}$$

Here, observer gain, state-feedback gain, and system state estimations are represented by L, K, and $\hat{x}$, respectively. The controller will now be implemented using the ANFIS-based nonlinear adaptive control design approach using CKF as an observer for the estimation of the system state.

The suggested design strategy is split into two sections:

(1) A CKF is being established to assess the instantaneous value of load power for uncertain and dynamic CPLs.

(2) Create an adaptive controller to optimise the injected current value of the energy storage unit.

Figure 4.5 depicts the suggested approach's control architecture. As shown in Fig.4.5, the CKF process is used to calculate P_{Load} in the DC microgrid. The proposed controller afterward applies this value, along with the iL, Vc values to calculate the suitable duty cycle for the switch, u.

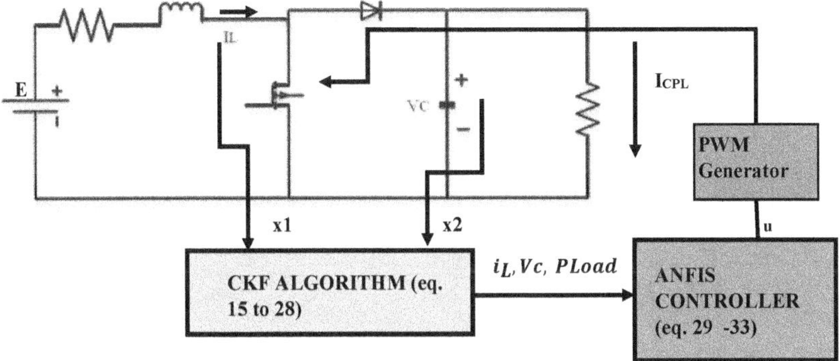

Fig.4.5: Proposed Controller Design Schematics

As the proposed approach is broken into two components, the mathematical derivation is done independently. The design must operate in parallel to employ the suggested controller. For the suggested control design apprach, the following algorithm is proposed:

(1) The power value is taken as P(0) initially

(2) Determine the DC microgrid's current and voltage values.

(3) Using the CKF technique, update the CPL power (4.12-4.25)

(4) Using an ANFIS-based algorithm, calculate the estimated current (4.35-4.38)

(5) The DC microgrid is fed with the estimated current.

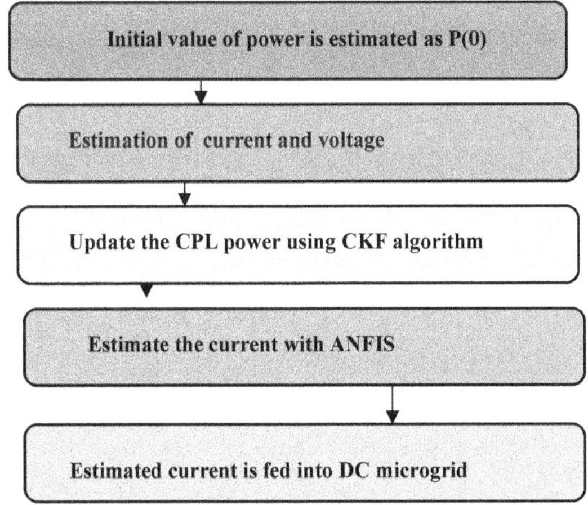

Fig.4.6: Control Flow diagram

Fig.4.6 displays the DC microgrid control flow diagram using the ANFIS algorithm.

4.2.3   Stability Analysis:

4.2.3.1 Stability analysis for the proposed method:

When renewable energy sources are increasingly incorporated into the grid and are connected to more converters, their dynamics and control have a major impact on the microgrid stability. Thus, it is essential to do a stability analysis at steady-state and during transients to provide a consistent and stable power supply. Lyapunov-based stability criteria are used in the execution of the stability analysis.

This method maximises an equilibrium's region of attraction (ROA) by algorithmically computing a Lyapunov function using the Takagi-Sugeno (TS) approach. Using state-space equations, the sources' dynamics are modelled dynamically. A Lyapunov function $V(x)$, fulfilling the Lyapunov conditions in an equilibrium region (where $x = 0$ and represents a state variable), is used to establish the stability requirements.

Considering the non-linear system $\dot{x} = f(x)$. If a Lyapunov function $V(x)$ exists that meets the requirements for Lyapunov criteria, the system will provide asymptotic stability$\left(\lim_{t \to \infty} x(t) = 0, \forall x_0 \neq 0 \right)$. $V(x)$ must be determined to test the system stability and is calculated as follows:

$$\dot{V}(x) = x^T(A^T M + MA)x \tag{4.47}$$

To support the system stability, $\dot{V}(x)$ is considered to be negative (less than zero). This happens only if the condition

$$x^T(A^T + MA)x < 0 \tag{4.48}$$

Therefore, the terms to establish stability with the help of Lyapunov function can be stated in linear matrix inequality (LMI) form and written as:

$$\begin{cases} M = M^T > 0 \\ A_i^T.M + M.A_i < 0 \ \ for \ \forall i = \{0,1,2,....r\} \end{cases} \tag{4.49}$$

The system's asymptotic stability is ensured by LMI (41)'s viability. The TS model can be used to accurately characterise the nonlinearities present in the systems under consideration in this work. Assume the non-linear model with n nonlinearities, $f_1(x)$ to $f_n(x)$. A sufficient number of local linear models can be used to represent the equivalent non-linear TS model, where every single nonlinearity $(f_i(x), i = 1,2,...,n)$ is defined by two values: $f_{i\ min}$ and $f_{i\ max}$ relating to a x_{imin} and a x_{imax}.

Only the *Ai* (state matrices), derived using the non-linear model, are required to show the stability (min or max value). The operational point's DA is calculated based on the region enclosed by the aforementioned boundary values. Plotted in Fig. 4.47 is the estimated domain of attraction for the proposed approach and traditional PI-based control.

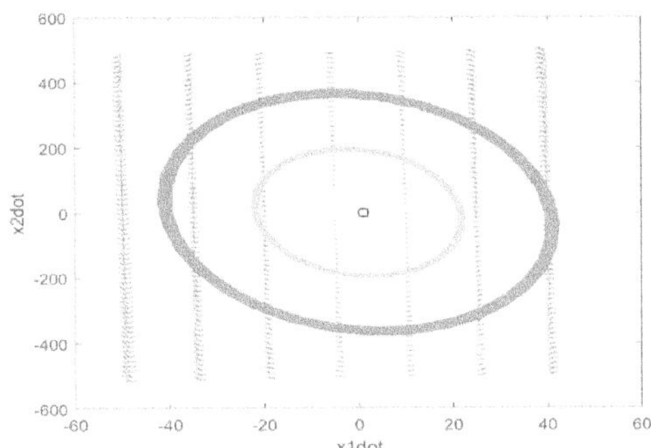

Fig.4.7: Assessed DA for the proposed control design(red) and traditional PI control design (green)

The suggested method has a bigger stability region than a conventional PI-based controller, as demonstrated in Fig.4.7.

4.2.3.2 Convergence Analysis for CKF:

The CKF's consistency directly impacts the DC microgrid's overall stability. As a result, a stability study demonstrating convergence is required for the estimation error to approach zero. [78], [79] proposed using the CKF algorithm to analyse the stability of nonlinear dynamical systems. This unit describes a process for analysing the CKF's convergence evaluation using the Lyapunov function. Estimation and prediction errors are defined as follows:

$$\begin{cases} \tilde{x}_{j+1} = x_{j+1} - \hat{x}_{j+1} \\ \tilde{x}_{j+1|j} = x_{j+1} - \hat{x}_{j+1|j} \end{cases} \tag{4.50}$$

Expanding x_j by a Taylor series about $\hat{x}_{j+1}$ gives,

$$x_j = f(\hat{x}_{j-1})(+\nabla f(\hat{x}_{j-1})\tilde{x}_{j-1} + \cdots + w_j \tag{4.51}$$

The following are simplified error expressions:

$$\tilde{x}_{j+1|j} \cong F_{j+1}\tilde{x}_j + w_j \tag{4.52}$$

Where

$F_{j+1} = (\frac{\delta f(x_j, u_j)}{\delta x_j}|x = \tilde{x}_j$, $x_j \in R^n$ is a state vector, $u_j \in R^n$ is the control input, and w_j is the process noise with covariance Q_j. By picking an unknown diagonal matrix β_j as follows, the model errors and the noise sequence w_j can be transformed into equality equations:

$$\tilde{x}_{j+1|j} = \beta_{j+1}F_{j+1}\tilde{x}_j \tag{4.53}$$

The Lyapunov function can be used to verify the asymptotic convergence of estimated inaccuracy determined with CKF for a nonlinear system provided by (3.7). If the following conditions apply, the CKF is asymptotically convergent locally [80].

Condition 1: Representing the system described by (3.7) to be equivalently observable, the resulting equality must be valid:

$$rank \begin{bmatrix} H_j \\ H_j \frac{\partial f}{\partial x}(x) \\ \vdots \\ H_j \left(\frac{\partial f}{\partial x}(x)\right)^{n-1} \end{bmatrix}_{x=\hat{x}_{j|j}} = n \tag{4.54}$$

Condition 2: The Matrix $Y_{j+1} = \frac{\partial y}{\partial x}|_{x=\hat{x}_{j|j}}$ is consistently bounded, also matrix Y_{j+1}^{-1} exist.

Condition 3: α_{j+1} , Q_j -the instrumental matrix and process noise covariance should be chosen to satisfy the condition :

$$\bar{\sigma}[\alpha_{j+1}]^2 \leq \underline{\sigma}[\alpha_{j+1}]^2 \frac{\underline{\sigma}[H_{j+1}]^2 \underline{\sigma}[P_{j+1|j}]}{\bar{\sigma}[P_{zz,j+1|j}]} + (1-\lambda)\frac{\underline{\sigma}[P_{j|j}^{-1}]\underline{\sigma}[P_{j+1|j}]}{\bar{\sigma}[Y_{j+1}]^2} \tag{4.55}$$

Where $\underline{\sigma}, \bar{\sigma}$ are the lowest and highest singular values, respectively, and $0 < \lambda < 1$. The following section supports the proposed CKF's postulates as mentioned above, theoretically confirming the nonlinear CKF's convergence.

4.3 Simulation and result :

This section reports the outcomes of the suggested adaptive ANFIS-based controller. The system model displayed in Fig.4.1 is used for simulation in Matlab/Simulink with the suggested controller displayed in Fig.4.5 to validate the efficacy of the suggested controller. The ANFIS controller is connected to the CKF approach, which determines the load's unidentified power during the period of time. In Table 4.1, the system parameters are provided.

TABLE -4.1

System Parameters with Various CPLs used in Simulation

Ls = 17.3 mH	*Cs = 1.05 mF*	*E (input) = 60.4(at maximum power point)* *Vdc = 200V*	*Rs = 0.4Ω*
L1 = 40 mH	*C1 = 1 mF*	*P1 = 900W*	*r1 = 0.8Ω*
L2 = 19.5 mH	*C2 = 1.05*	*P2 = 600 W*	*r2 = 0.42Ω*

Two situations are shown taking the assistance of MATLAB/SIMULINK.

Situation-1: the proposed controller architecture compares the error outputs for a given parameters (Table 4.1) with the help of direct estimation method and the CKF algorithm. In both cases, the estimated error is shown in Fig.4.8. The proposed control approach and the CKF algorithm are combined in situation-2 to give the DC microgrid fast transient performance and robustness both, and the effectiveness of the proposed design is proven.

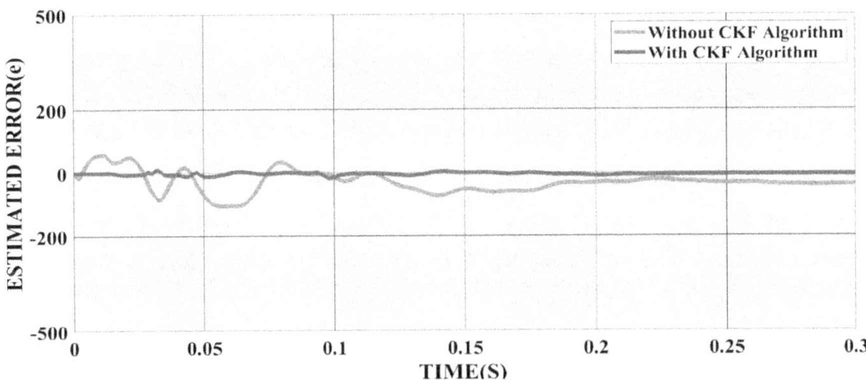

Fig. 4.8: Comparison of DC Microgrid estimation errors (Proposed control approach with direct estimation and using CKF algorithm)

Scenario-1 :

By setting the initial energy storage current to zero, the DC microgrid characteristics indicated in Table 4.1 should guarantee the system's stability by achieving its equilibrium point during disturbances.

The DC microgrid's preliminary condition is set at $X0 = [7.5\ 200\ 7.5\ 200]^T$. The noise variances for the system and measurement are $w = 0.001I$ and $v = 0.01I$, respectively.

Initially, the p value is chosen as a diagonal member of the given matrix, which is linked to the expected variance of the next state. Because i_L and v_C are assessable, and $PLoad$ is unidentified, $p0 = diag(1,1,1,103)$ is chosen as the primary value of the covariance matrix p. The size of the covariance matrix is set to allow for speedy convergence. In Fig .4.8, the assessment errors with CKF, and direct method are drawn and compared, and the fig. 4.8 shows that utilising CKF delivers better results. When using the CKF, the estimated errors approaches near zero after a small period (about 0.15 s), however, the assessment errors fluctuates throughout the test time with direct method. Without CKF, the assessed values do not approach to zero.

Scenario-2:

The working of the DC microgrid connected to renewable energy sources faces the critical issue of high-power transients. The CPL's nonlinearity drives the microgrid towards an unstable region during transients.

The design approach should guarantee consistent performance even at transients while ensuring system stability. Transients produce deviations in the CPLs voltages and currents, which disrupts the equilibrium condition of the system.

The simulation outcome is presented in two scenarios: when transients happen and when a immediate load change happens. A MATLAB simulation is run to test the efficacy of the suggested method, and the outcomes are estimated and compared with the designs of fuzzy-logic and PI-based controllers.

The simulation is run with the system constraints given in Table-3.1 as input. Figure 4.9 -4.14 demonstrates the simulation findings for the proposed technique, fuzzy-logic control, and PI control in contrast to the functioning of the DC microgrid.

Case 1: During transients, the capacitor voltages at CPL1 and CPL2 are displayed in Figs. 4.9 and 4.10. The outputs of the suggested ANFIS controller design are compared for the PI and fuzzy-logic designs. The value of current injection via ESS is compared for all three approaches during transients also, as illustrated in Fig. 4.13.

Case 2: Next, output voltage changes are considered, and system outputs are calculated using simulations and compared. In this case, CPL is kept at 900 W.

When a 600 W additional load is supplied to the system, the load demand changes rapidly in 0.5 seconds. Figures 4.11 and 4.12 explain the variance in output during load change for the given CPLs and the reactions of the fuzzy-logic and PI-based controllers.

The ESS currents at the time of load variation for all three approaches are also compared and shown in Fig.4.14.

Compared to the standard PI-based and the fuzzy-logic controller, the CKF-based suggested ANFIS control approach outperforms. As shown in Fig.4.9 -4.14, the proposed controller enhances transient and steady-state performance and utilizes less injecting current than previous approaches to stabilise the DC microgrid as well.

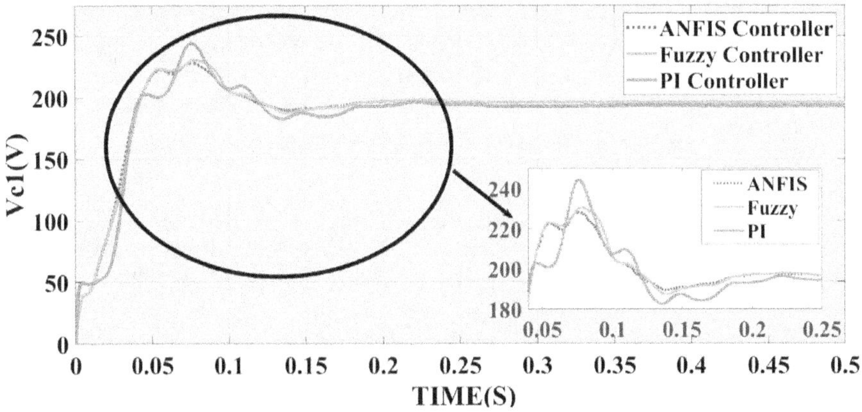

Fig.4.9: Voltage during transients at CPL1

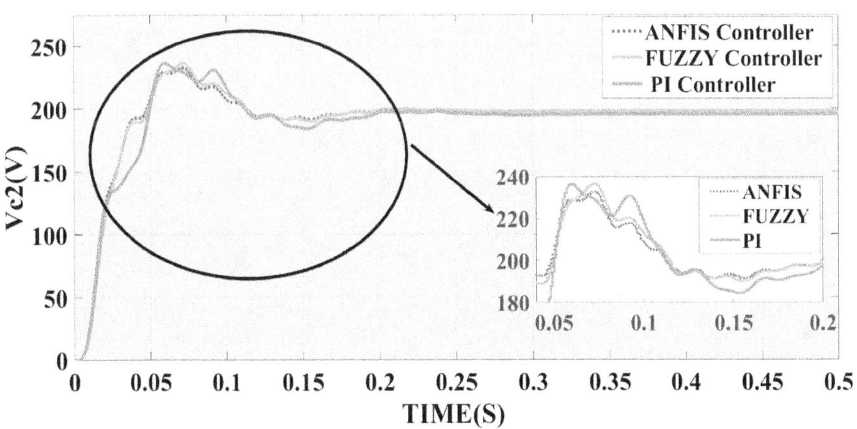

Fig.4.10: Voltage during transients at CPL2

68

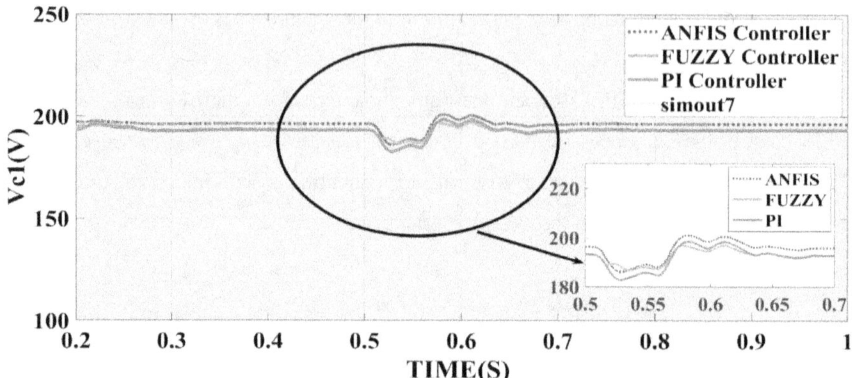

Fig.4.11: Voltage at CPL1 during Load change

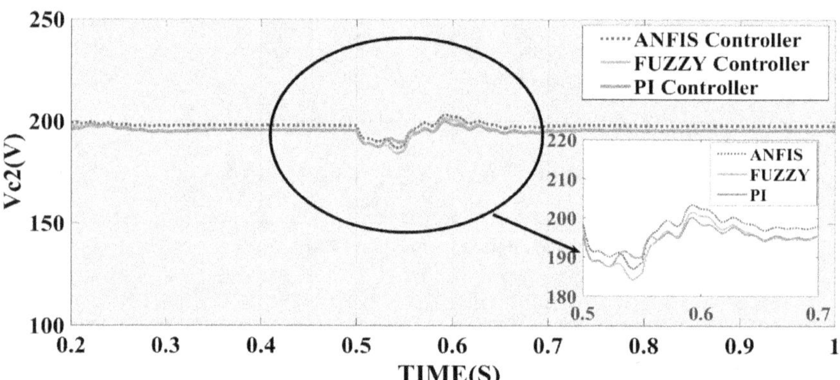

Fig.4.12: Voltage at CPL2 during Load change

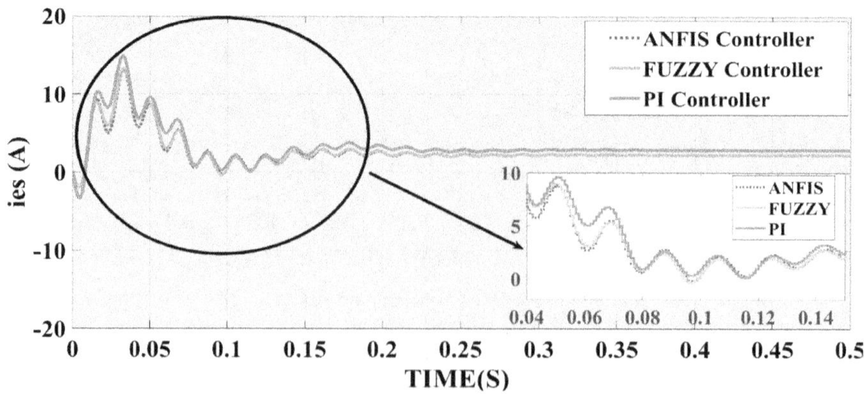

Fig.4.13: ESS Current (ies) during transients

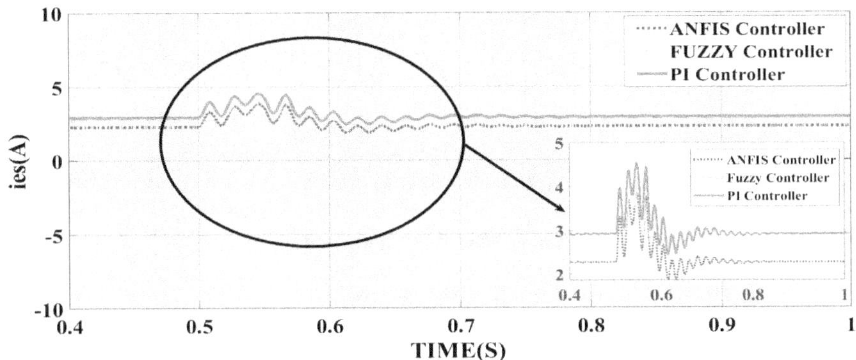

Fig. 4.14: ESS Current (ies) during load change

Established on simulations in Matlab/Simulink, Table -4.2 has been included to assess the settling time for bus voltage, norm of control signals, and overshoots for the other design approaches referred above. Generally, the feedback matrix's norm must be kept as small as feasible to minimise the transient response for a closed-loop system [81].

Table 4.2 compares the control signal norm derived using MATLAB for ANFIS, PI-based, and fuzzy-logic control designs. In comparison to traditional PI and fuzzy-logic design approaches, the ANFIS solution has the lowest control signal norm. As a result, in comparison with fuzzy-logic and PI-based approach, the ANFIS architecture exhibits a reduced percentage overshoot. In addition, compared to PI and fuzzy controller designs, the ANFIS approach delivers the quickest settling time after the transients, as shown in Table 4.2.

TABLE -4.2

The evaluation of the performance using different methods for DC microgrid Control

Controller Design Method	Settling time(s)	Percentage overshoot	Norm of control
PI Control	*0.50*	*0.25*	*1.9847*
Fuzzy -logic control	*0.25*	*0.15*	*0.4337*
Suggested design(CKF+ANFIS)	*0.2*	*0.14*	*0.2628*

Estimated error Stability Verification (CKF Method):

In order to demonstrate stability, the nonlinear system (4.41) must meet the requirements stated in (4.48) and (4.49) by solving the equation given below:

$$rank\begin{bmatrix} H_k \\ H_k & F_{k+1} \end{bmatrix}_{x=\hat{x}_{k|k}} = rank\begin{bmatrix} 1 & 0 & 0 & 0 \\ 0 & 0 & 1 & 0 \\ \frac{-r1}{L1} & \frac{-1}{L1} & 0 & \frac{1}{L1} \\ 0 & 0 & \frac{-r2}{L2} & \frac{-1}{L2} \end{bmatrix} = 4$$

As a result, condition I can be verified because the approach is consistently observable.

To verify condition II, the matrix Z needs to be computed.

$$Z = \begin{bmatrix} -27.85 & -25.316 & 0 & 25.316 \\ 1996 & 554,400/x^2 & 0 & 0 \\ 0 & 0 & -27.848 & -25.316 \\ -1814 & 0 & 1814 & 0 \end{bmatrix}$$

Z is uniformly confined since Z^{-1} exists and x_2 occurs inside a positive interval close to the operational point. This confirms an assumption from II. Moreover, matrix $\beta_k = 0.001I$ and the covariance $Qk = 0.001I$ provide assurance for Assumption III.

4.4 Conclusions :

This chapter outlines a method for designing an observer-based DC microgrid controller. The system parametric uncertainties are taken into account in the control problem formulation as system uncertainties due to faulty knowledge of grid inductance and impedance.

The suggested observer-based controller is intended to compensate for grid voltage fluctuations while managing a good performance. Because the controller relied on observer-generated states, the number of measurements necessary was significantly reduced, lowering the implementation cost. Various case studies are used to explain the efficacy of the suggested control strategy. The presented outcomes show that the suggested controller's observer and control design performance all meet the design requirements.

This section shows how an ANFIS-based adaptive controller using a CKF algorithm controls the current from an ESS connected to a DC microgrid with several CPLs. The controller design is developed by treating all component parameters as completely unknown and then adapting them using adaptation laws. This work aims to stabilise a DC microgrid with unpredictable dynamic loads. The CKF calculates the instantaneous value of the uncertain load power in augmented states. After that, the estimated state is used to control the system through the application of the suggested controller design approach.

The suggested method is resistant to uncertainty and has a short calculation time. Because of the asymptotic stability criteria and the CKF's speedy convergence estimate, a controller with better transient outcomes is produced. To demonstrate the effectiveness of the suggested controller, two situations are studied.

In scenario 1, the CKF algorithm's effectiveness is tested and contrasted. The error signal is generated using the CKF technique and without it. The comparison indicates that when the CKF algorithm is applied, the error signal convergence is faster than when it is not. The stability criteria of the CKF algorithm are also examined. All through the transient conditions and during load-change, the effectiveness of the design approach using the CKF algorithm is verified in scenario-2.

The suggested controller's transient response is assessed and compared with the fuzzy-logic and PI-based design approach. The control norms, settling times alongwith percentage overshoots are also compared. In comparison to the other approaches, the suggested control approach has the least overshoot and the quickest settling time.

The control method is also tested and confirmed in the event of a sudden load shift. After 0.5 seconds, the load is abruptly raised, and the voltage is assessed and assessed in comparison with the fuzzy-logic and PI-based control design outputs. When combined with the CKF algorithm, the suggested controller assesses the system state (inductor current and capacitor voltage) and reduces transient overshoots. After transients, the suggested controller reduces the settling time to a minimum. In comparison to PI controller and fuzzy-logic controller design, the comparison ensures the efficacy of the suggested strategy.

The norms for the ANFIS controller using the CKF guarantee the closed-loop stability for the DC microgrid. The stability performances have been confirmed and are included in the appendix. Investigating the DC microgrids' stability when coupled to an AC microgrid may be an excellent future study topic. It is also suggested that the proposed technique be used for multiple microgrid setups with multiple loads, different grid topologies, and renewable sources.

CHAPTER 5

ADAPTIVE ROBUST CONTROL FRAMEWORK:

5. Adaptive Robust Framework for CPLs connected DC Microgrid:

Practically every physical structure has some degree of model uncertainty, making the development of control algorithms with high performance a difficult task. In most cases, the origins of model uncertainty are divided into two groups:

(i)   Unknown physical parameters that are repetitive or consistent.

(ii)   Unidentified quantities, for example, external disruptions and inaccurate modelling of some physical variables which are non-repetitive

To account for these uncertainties, two nonlinear control strategies—deterministic robust control (DRC) [82]–[85] and adaptive control [86]–[88] may be utilised. Deterministic robust controllers can achieve assured transient and final tracking accuracy during both uncertain parameters and unknown nonlinearities. The design approach, however, is conservative and impractical because no effort is made to draw lessons from past behaviour to minimize the consequences of constant unknown elements. Conversely, adaptive controllers can accomplish asymptotic tracking even with parametric uncertainty [89].

Yet, in addition to parametric uncertainties and unknown nonlinearities, some systems are also susceptible to dynamic uncertainties that depend on the unmeasured states of external dynamic systems. Parameter uncertainty consists of testing errors, measurement errors, unpredictability, and alternate data use.

This chapter discusses an adaptive robust control technique for nonlinear CPLs that are subject to dynamic uncertainties, unknown nonlinearities, and parametric uncertainties. In general, adding disturbances to a controller reduces both transient and steady-state performances.

The addition of disturbances for an adaptive approach based on tracking error may result in the exponential expansion of the parameter estimates. The presence of disturbances worsens the problems for control systems that use a prediction error-based update strategy because the prediction error is calculated using a filtering mechanism. When applying the filtering method to dynamics with external disturbances, the unidentified disturbance terms are simultaneously filtered and integrated with the filtered control input, which is undesirable. When implemented on the actual plant, the ultimate purpose of any control design is to achieve good performance during external disturbances. To put it another way, the controller should ensure closed-loop stability and appropriate performance for the nominal plant model and a family of plants, which will almost certainly include the actual plant. Nonlinear adaptive control can reduce uncertainties by adjusting parameters to attain asymptotic stability, whereas deterministic robust control applies robust feedback to limit the impact of model uncertainty. The limitations of adaptive control include the requirement for certain invariant properties, such as unknown parameters assumed to be constant.

Furthermore, the presence of small measurement noises and disturbances can cause instability. On the other hand, robust control design lacks the ability to reduce tracking errors. CPLs in DC microgrids tend to make the system unstable by lowering the effective damping. As a result, it is difficult to operate and regulate the converters, which seriously disrupts the DC microgrid's transmission system. Therefore, a suitable control framework is needed to minimise the destabilising consequences caused by CPLs.

This chapter presents an intelligent control that provides a robust and adaptive framework to control the nonlinear disturbances caused by CPLs.

5.1 General Framework of Robust Adaptive Control :

This chapter presents a common framework for the adaptive robust control used for a nonlinear system.

5.1.1   Problem Statement:

Considering the following general equations for the nonlinear system:

$$\begin{cases} \dot{x} = f(x,\theta,t) + B(x,\theta,t)u + D(x,t)\gamma(x,\theta,u,t) \\ \qquad\qquad y = h(x,t) \end{cases} \tag{5.1}$$

Where, $u \in R^n$ and $y \in R^n$ are the input and output vectors, respectively. $x \in R^p$ represents the state vectors and $\theta \in R^m$ represents the unknown parameters vectors. $h(x,t)$, $f(x,\theta,t)$, $B(x,\theta,t)$ and $D(x,t) \in R^{pxl}$ are known vectors, while $\gamma(x,\theta,u,t)$ denotes the unknown nonlinear function vectors such as modelling errors or disturbances. The following assumptions we make for uncertainties in a nonlinear system:

Assumption 1: The unknown nonlinear functions, along with parametric uncertainties comply with the following:

$$\begin{cases} \qquad \theta \in \Omega_\theta \triangleq \{\theta : \theta_{min} < \theta < \theta_{max}\} \\ (x,\theta,u,t) \in \Omega_\gamma \triangleq \{\gamma : \|\gamma(x,\theta,u,t)\| \leq \epsilon(x,t)\} \end{cases} \tag{5.2}$$

Here, θ_{min}, θ_{max} and $\varepsilon(x, t)$ are known.

Let $y_k(t) \in R^n$ represent the needed output at t. Additionally, the tracking errors be denoted as $e_y = y - y_k(t)$. The challenge of creating a control approach for the input u is such that the output tracking provides a required transient performance and accuracy with global system stability under the assumption of (5.2). Besides that, asymptotic output tracking should be possible even in the presence of parametric uncertainties.

5.1.1.1 State Observer using CKF with H-infinity error bound:

The observation of state variables is one of the core issues with dynamic systems. The state estimation with a worst-case dynamic system design aim has not been taken into consideration in earlier research, despite the quadratic error criterion having a strong theoretical base. All theoretical results are established on the theory that the system model is perfectly defined. However, model uncertainty and incomplete statistical information are frequently encountered in real-world applications, leading to excessive estimation errors.

A robust nonlinear observer is needed to address these issues to reduce the interference caused by unspecified input errors, unpredictable interference, or faults. Examples of nonlinear observers contain the unscented Kalman filter (UKF), the cubature Kalman filter(CKF), and the extended Kalman filter (EKF).

The weight can always be maintained at positive in the CKF algorithm irrespective of the state dimension because it does not require any additional parameters for the sampling operation. This allows the filtering to proceed without interruption[90]. As a result, CKF is more effective than other filtering techniques for state estimation in high-dimensional systems. However, the standard CKF algorithm loses its symmetry and positive definiteness over repeated iterations due to the accumulation of rounding errors.

Recent years have seen extensive research on this phenomenon. An adaptive Huber algorithm consisting of several robust tracking is incorporated into the standard CKF in[91]. This can significantly increase the CKF's filtering performance.

In [92], an adaptive component based on outstanding prediction is being developed to reduce the abnormal effects caused by CKF interference model errors, such as an error in state estimation. The system state covariance matrix is decomposed and iterated using the QR decomposition method, which has good numerical stability to decrease the complexity during filtering operations [93]. A state estimation technique called the H-infinity method is predicated on the concept that the worst possible noise interference situation exists. The approach was developed to reduce the impact of the worst disturbances on estimation errors. It can be updated to accommodate faulty system modelling, making it useful for the state estimation problem when the system's measurement noise's statistical features are unknown [94]–[96]. Yet, the change in noise's statistical properties decreases the H-infinity technique's accuracy.

Therefore, this chapter combines the H-infinity technique with the CKF to increase the nonlinear filter's accuracy. The H-infinity Cubature Kalman Filter (HCKF) is presented in this chapter to maintain the benefits of both the CKF and the H-infinity method and to ensure strong robustness during large disturbances in the system.

The disturbances can be due to sudden and large load changes, external faults, and initial transients. HCKF increases the system's robustness and reduces estimation inaccuracy in the event of interference. It ensures that the filter can continue to function normally in the event of significant unexpected noise and has a higher filtering accuracy.

The general framework for the filtering problem is presented below:

$$\begin{cases} x_j = x_{j-1} + Bu_{j-1} + w_{j-1} \\ \quad\ z_j = H_j x_j + v_j \end{cases} \tag{5.3}$$

In the equation above, B represents the input matrix for control. The system's state vector is represented by $x_j \in R^n$. z_j is the measurement vector. $H(.)$ denotes a nonlinear function. w_{j-1} and v_j represent system noise and measurement noise, correspondingly. Noise terms w_{j-1} and v_j can be unknown or deterministic.

The concept of H-infinity estimation is based on finding the method for cost function minimisation when P_0, Q_j, and R_j attain the upper bound. Here, Q_j, and R_j indicate the covariance matrices of w_j and v_j for the state and measurement, respectively. P_0 stands for the initial estimation error covariance matrix, which indicates how near x_0 is to the initial estimate $\hat{x}_0$. The matrix P_0 is set up following specific problems. An algorithm that combines the cubature Kalman filter and the H-infinity filter is suggested in this chapter for state estimation of the nonlinear system.

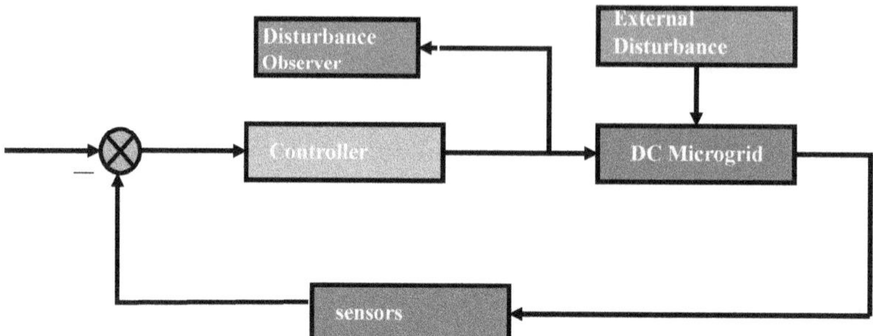

Fig.5.1: Block diagram of an Adaptive Robust Controller

A simple block diagram of the robust adaptive method is displayed in Fig. 5.1.The H-infinity Cubature filter (HCKF) algorithm [97] can be summed up as follows:

5.1.1.2 H-infinity Estimation fundamentals:

H-infinity filtering presents the concept of the H-infinity norm for the ambiguity of the system model and noise data characteristics. The H-infinity norm from the interference input to the error output of the filtering must be maintained to a minimum to reduce the estimated system error in the worst instance of interference.

$$H_\infty = \frac{\sum_{j=0}^{n-1}\|Y_j - \hat{Y}_j\|_{D_j}^2}{\|X_0 - \hat{X}_0\|_{P_0^{-1}}^2 + \sum_{j=0}^{n-1}\left(\|w_j\|_{Q_j^{-1}}^2 + \|v_j\|_{R_j^{-1}}^2\right)} \tag{5.4}$$

The H-infinity estimator's primary goal is to reduce state estimation inaccuracy. In the worst scenario, w_j, v_j and X_0 shape the boundary of H-infinity and provide a threshold value:

$$sup H_\infty < \gamma^2 \tag{5.5}$$

Where sup expresses the maximum value of the cost function, and γ is the attenuation limit for errors. The aim is to find $\hat{X}_j$ so that $H_\infty < \gamma^2$ can include any interference introduced by w_j, v_j and X_0 and also minimises H_∞ in the worst situation. The H-infinity estimator challenge can be solved by answering the following inequality equation known as the Riccati inequality equation[20]:

$$P_{j+1}^{-1} = P_j^{-1} + \nabla H^T \nabla H - \gamma^2 I_p > 0 \tag{5.6}$$

In the above equation, P_k denotes the error covariance matrix, and I_p denotes the p -order identity matrix. At the predicted value of $\hat{x}_{k|k-1}$, ∇H is the Jacobian matrix expanded by $H(.)$.

The error constrains the parameter γ's estimated error for the H-infinity estimation. The robustness of the system is higher when the γ is less. However, a value that is too small may cause filter divergence. Conversely, as γ increases, it will decrease the robustness of the system.

5.1.1.3 Cubature H-infinity Observer:

The H-infinity Cubature Kalman filter (HCKF) is obtained by combining the Cubature Kalman filter algorithm with the H-infinity estimation method. The time prediction stage in the HCKF algorithm includes state prediction, estimating Cubature points propagation, and factor decomposition of the error auxiliary matrix similar to CKF. In the measurement update step, the covariance matrix and updated state of the H-infinity estimator can be represented as written below:

$$
\begin{cases}
H_{\infty,j} = P_{j|j-1}\nabla H^T\left(I + \nabla H P_{j|j-1}\nabla H^T\right)^T \\
\quad \hat{x}_j = \hat{x}_{j|j-1} + H_{\infty,j}\left[Z_j - \hat{Z}_{j|j-1}\right], \\
P_j = P_{j|j-1} - P_{j|j-1}[\nabla H^T I]R_j^{-1}\begin{bmatrix}\nabla H \\ I\end{bmatrix}P_{j|j-1}^T
\end{cases}
\tag{5.7}
$$

Where, R_j used in the equation above represents the assessed value of the observation noise. Liang et al. [98] provided the mathematical-analytical formula for the assessed value of the observation disturbance depending on the sample value:

$$
\hat{R}_j = \begin{bmatrix} I & 0 \\ 0 & -\gamma^2 I \end{bmatrix} + \begin{bmatrix}\nabla H \\ I\end{bmatrix}P_{j|j-1}[\nabla H^T \quad I]
\tag{5.8}
$$

In [99], the error covariance and cross-covariance calculation techniques using the Jacobian matrix are presented as:

$$
\begin{cases}
P_{xz,j|j-1} \approx P_{j|j-1}\nabla H^T \\
P_{zz,j|j-1} \approx \nabla H P_{j|j-1}
\end{cases}
\tag{5.9}
$$

Formula (5.8) transforms formula (5.9) as follows:

$$
\hat{R}_j = \begin{bmatrix} I + P_{zz,j|j-1} & P_{zz,j|j-1}P^T_{xz,j|j-1}(P_{xz,j|j-1}P^T_{xz,j|j-1})^{-1}P_{j|j-1} \\ P_{xz,j|j-1} & P_{j|j-1} - \gamma^2 I \end{bmatrix}
\tag{5.10}
$$

The covariance matrix can be updated as:

$$
P_j = P_{j|j-1} - [P_{xz,j|j-1} \quad P_{j|j-1}]\,\hat{R}_j^{-1}\begin{bmatrix}P^T_{xz,j|j-1} \\ P_{j|j-1}^T\end{bmatrix}
\tag{5.11}
$$

The updated equation for the gain can be obtained by inputting equation (5.11) into equation (5.7).

$$H_{\infty,j} = P_{xz,j|j-1}\left[P_{zz,j|j-1} + I\right]^{-1} \tag{5.12}$$

The noise error is computed using (5.8), and later Riccati inequality (5.6) is converted using (5.9). The HCKF can be thus obtained by incorporating the gain update, state estimation, and error covariance estimate into the filtering step of the CKF and expressed as:

$$P_{j+1}^{-1} = P_j^{-1} + P_{j|j-1}^{-1}P_{xz,j|j-1}P_{zz,j|j-1}P^T{}_{xz,j|j-1}\left(P_{xz,j|j-1}P^T{}_{xz,j|j-1}\right)^{-1} - \gamma^2 I_p > 0 \tag{5.13}$$

The measurement update calculation is performed built on the existing condition after choosing the suitable error limiting parameters. The gain of the HCKF filter is calculated using (5.12). The state value can be updated by referring to (5.7), and the error covariance matrix can be computed using (5.10)-(5.11).

The gain value and error covariance matrix computed using the H-infinity estimator are applied in place of K_j and P_j in the CKF algorithm. The CKF algorithm is generated by updating the measurement noise using the H-infinity estimation. The measurement updating section includes the H-infinity estimation theory to continually update and adjust the measurement noise and increase the filtering precision. Therefore, the HCKF algorithm improves the system's stability compared to the conventional CKF algorithm.

The flow chart is depicted in Fig.5.2 as shown below:

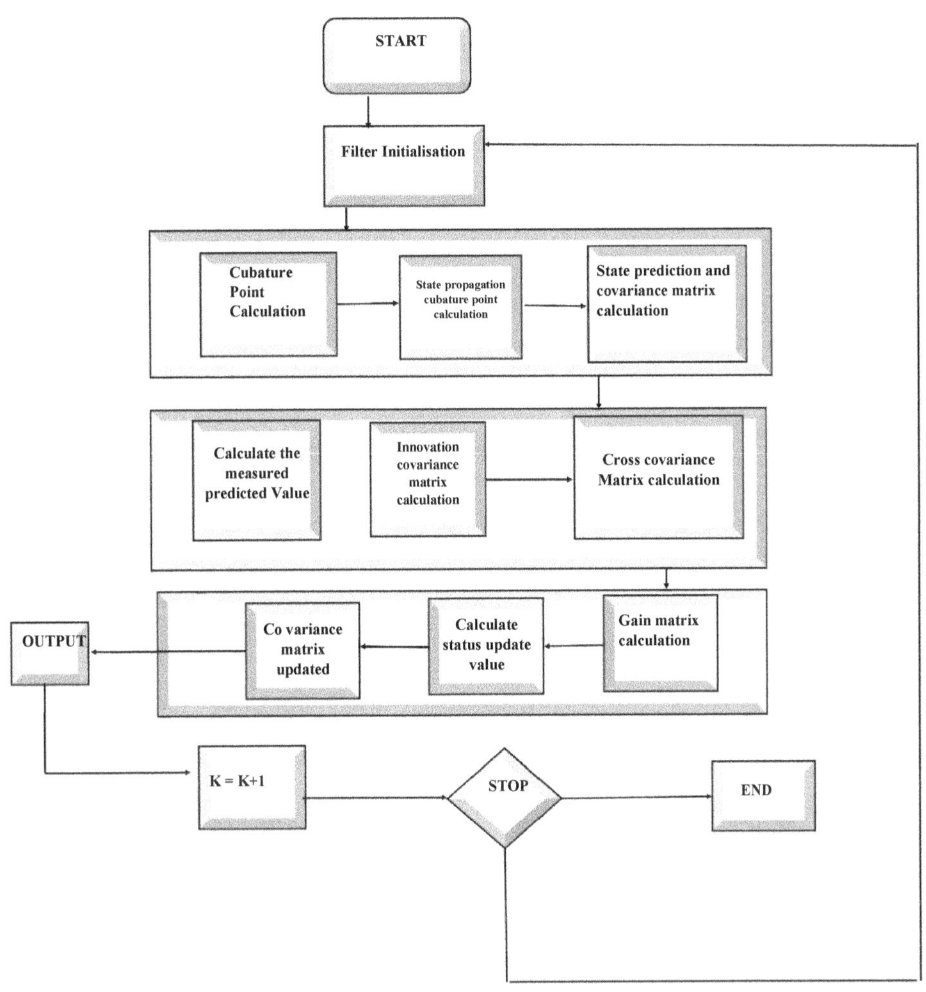

Fig.5.2: Flow Chart for HCKF

5.1.1.4 HCKF and CKF Comparison:

The convergence test and error estimations for HCKF and CKF are conducted using MATLAB/Simulink. The findings are assessed and matched to evaluate the approach's efficacy. A disturbance is introduced at 0.12 seconds to test how both algorithms behave. The HCKF-based algorithm converges more quickly for disturbance compared to the CKF algorithm, as shown in fig.3. Also, a comparison between the root mean square(RMS) errors of the CKF and HCKF during disturbance is shown in Fig. 5.3. According to the error displayed in Fig. 3, the RMS error of the HCKF algorithm is less compared to CKF algorithm. Table-1 shows the assessment of the calculated RMS error for both algorithms.

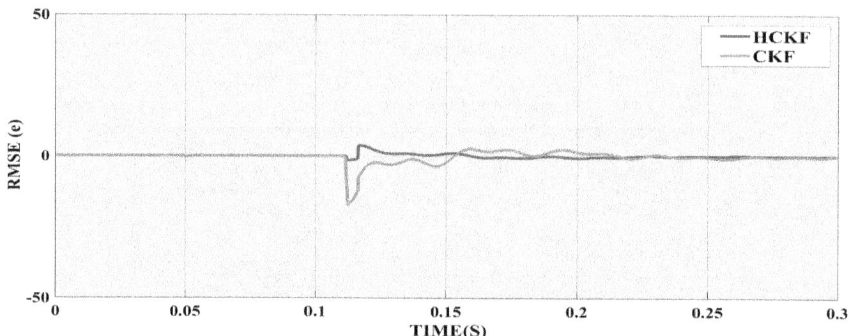

Fig.5.3: Comparison of RMSE for HCKF and CKF algorithm

TABLE-5.1

RMS Error comparison for HCKF and CKF algorithm

	CKF	HCKF
RMSE	0.7851	0.2857

The comparison in Table-5.1 shows that the estimation error during disturbance is less in HCKF compared to the CKF algorithm.

5.1.2   Intelligent Adaptation for ANFIS-based Controller:

Artificial intelligence (AI) is in use for various businesses and research domains. An "intelligent control" method uses various AI techniques, including machine learning, genetic algorithms, neural networks, and fuzzy logic. In this study, the design approach for the DC microgrid is implemented by applying the ANFIS.

The ANFIS system is a flexible, learner-based network with a shared fuzzy inference scheme. This system is based on the first-order Takagi-Sugeno-Kang (TSK) fuzzy inference system, which produces an initial fuzzy system before using learning techniques to control the constraints for making the error zero. The overall block configuration with ANFIS system is exhibited in Fig. 5.4 to demonstrate the system with the fewest errors possible.

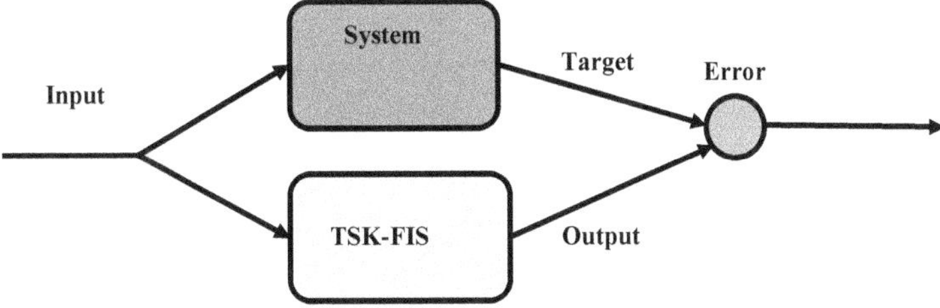

Fig.5.4: Simple block diagram of ANFIS-based system

The behaviour of the model can be brought closer to that of the underlying system by altering model parameters. To address this issue, an average square of the error-index is presented as a cost function.

$$Mean\ Square\ Error = \frac{1}{N}\sum_{i=1}^{N} e_i^2 \ \text{--------------------}(5.14)$$

e presents the degree of error, while N presents the number of data. The model's requirements ought to be changed to give the cost function the minimal achievable value. This situation is known as training. The ANFIS controller's working process was divided into two primary steps: training approaches are used to converge the values to a point where the error rate can be zero after constructing an initial fuzzy system.

The two learning criteria used by this method are the error produced relating the output voltage and the reference voltage for the load and the differentiation of this error. Simulations are run after the trained ANFIS model is integrated into the control system. The findings demonstrated that using ANFIS improved the simulation's ability to match experimental data.

5.2 Simulation and result

The proposed robust adaptive controller's findings employing the ANFIS-based control design are shown in this section. The controller design is combined with the HCKF algorithm, which calculates the load's power, which is unpredictable and varies over time. First, the HCKF is applied to evaluate *Pload*, the load's total power. The ANFIS-based controller is then sent the estimated Pload as an input. Fig. 5.5 displays a block diagram of the suggested strategy.

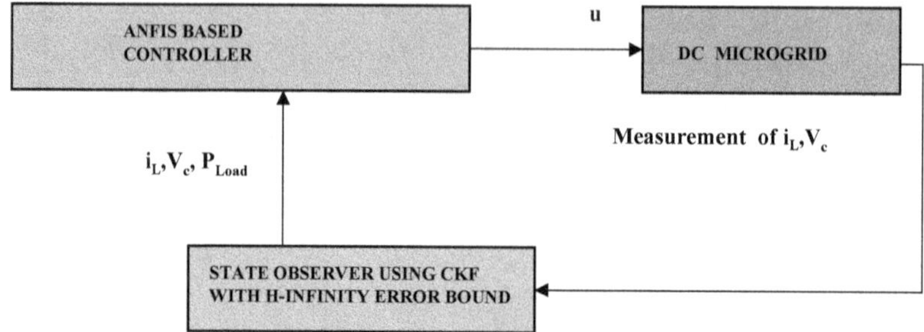

Fig.5.5: Block diagram for the proposed controller

Simulations using MATLAB/Simulink were done to evaluate the validity of the suggested approach. The case analysis is done under three different scenarios: (i) an Initial transient Scenario, (ii) a changing load scenario, and (iii) a fault scenario based on the criteria listed in Table 1 (chapter -3)

Two cases are given to demonstrate how well the suggested nonlinear controller works. In the first scenario, the entire load's power, Pload, suddenly increases to a high figure; in the second scenario, a fault is inserted later to assess the design's dependability. The outcomes of both scenarios' simulations are described in the following paragraphs. The HCKF method is utilised in both cases to evaluate the uncertain load power. The adaptive ANFIS-based controller then adjusted the DC microgrid's voltage to the intended reference of 200 V.

Scenario 1(Initial Transient Response):

The output voltage at initial transient conditions for the suggested controller is compared with the outputs of the traditional PI controller, H-infinity controller, and ANFIS-based Controller (without observer). The simulation response for voltage and current are displayed in Fig.5.6-5.9. Voltage outputs at CPL1 and CPL2 are demonstrated in Fig.5.6- 5.7. While inductor currents are presented in 5.8-5.9.

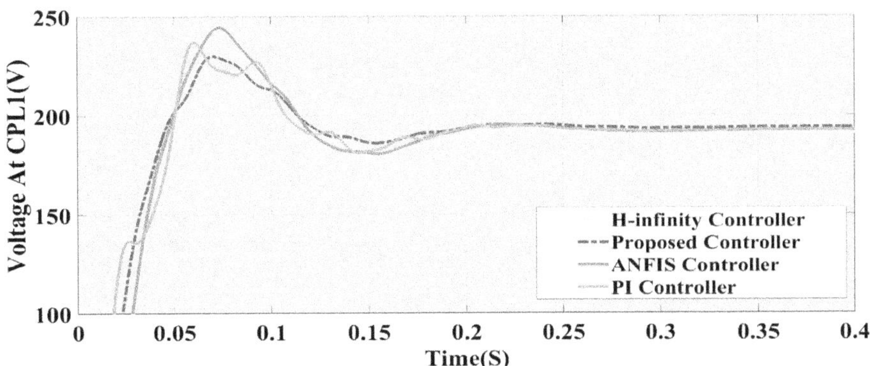

Fig.5.6: Voltage transient response at CPL1

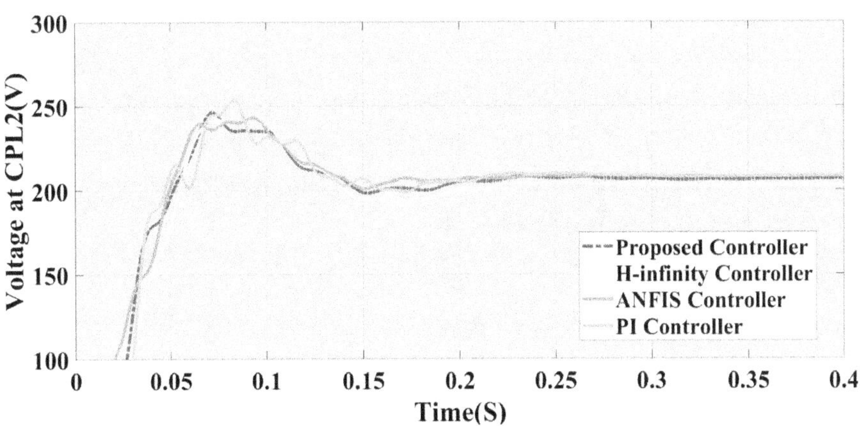

Fig.5.7: Voltage transient response at CPL2

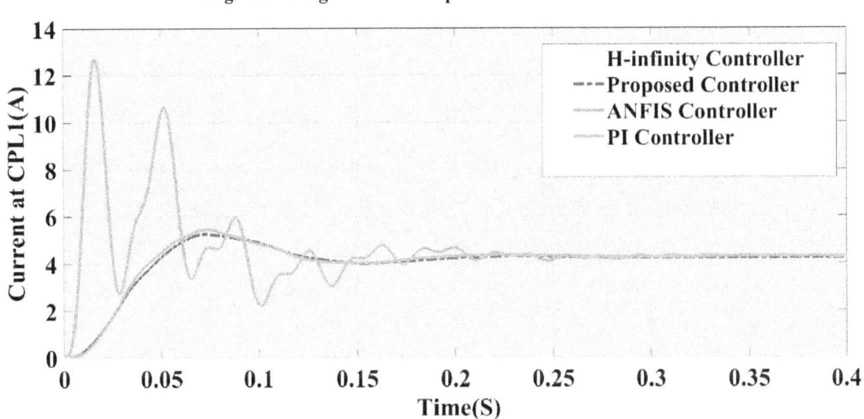

Fig.5.8: Current transient response at CPL1

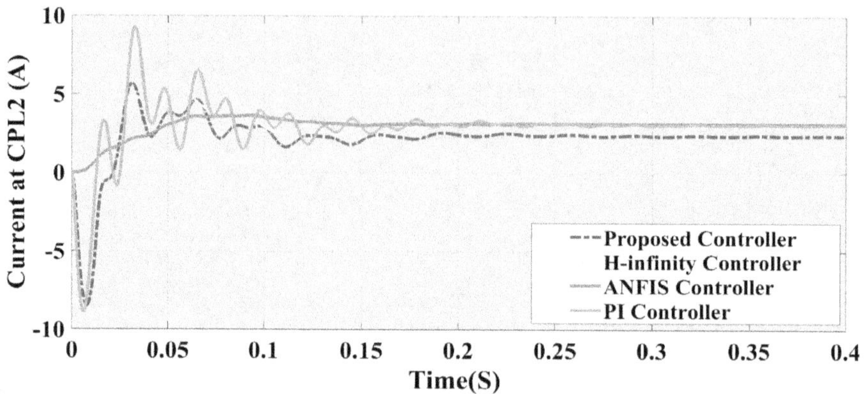

Fig.5.9: Current transient response at CPL2

Although the other controllers can stabilise the bus voltage, there is still a chance for an increase in transient responses, as indicated in the diagram. In comparison to the suggested controller, the overshoot is greater and takes more time to approach to steady state. According to the comparison, the suggested controller can provide improved transient output and minimises the percentage overshoot, including the settling time.

Scenario 2(Load change):

The PV source is operating at standard irradiance and temperature. The system receives an additional 600W load at time t = 0.5 sec. The requirement for load will go from 900W to 1500W. The output voltage can be significantly disturbed by a large change in the load. For comparative purposes, the findings of the suggested controller are assessed in comparison with a traditional PI controller, an ANFIS-based controller (without an observer), and an H-infinity controller. The simulation results for bus voltage and current response under varying load demand with all of the aforementioned controllers are shown in Fig. 5.10-5.11, respectively.

The bus voltage becomes unstable at time t = 0.5s as a result of the increased load demand. The diagram demonstrates that even with considerable CPL changes, the system may still achieve stability when employing the proposed controller design to correctly track the reference voltage at 200 V. Compared to other controllers, the system stabilises quickly. Also, the current value is low at the bus for the proposed controller compared to other controllers during load variations.

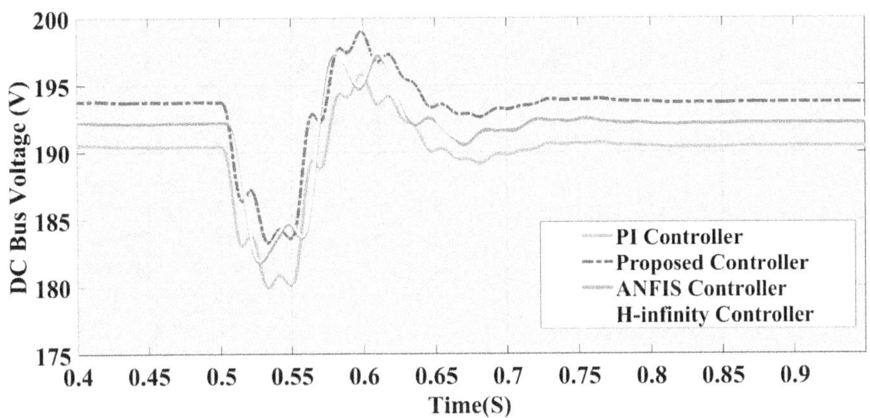

Fig.5.10: Bus Voltage during load change

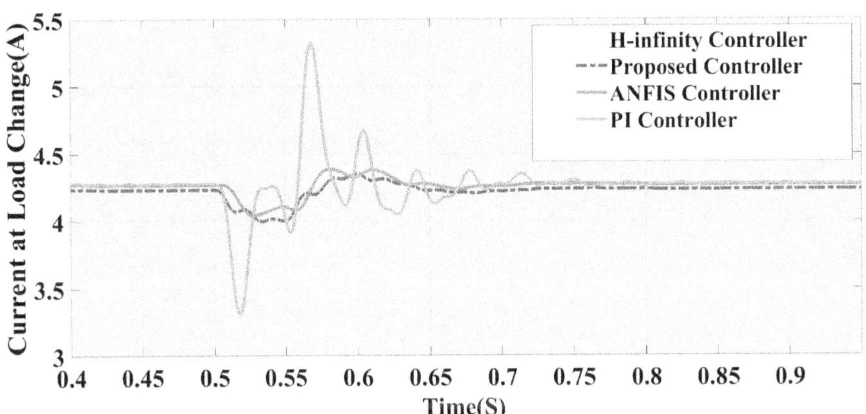

Fig.5.11: Bus Current during load change

Scenario 3(Fault condition):

To assess the effectiveness of the suggested controller, a fault is introduced at the DC link at 0.5s. As shown in Table -1, the system is operating at its nominal specifications. If the issue is not fixed, it will put the system in an unstable state. All of the controllers' DC bus voltage and current simulation results are compared. Figs. 5.12-5.13 show the simulation diagrams for the bus voltage and current, respectively.

86

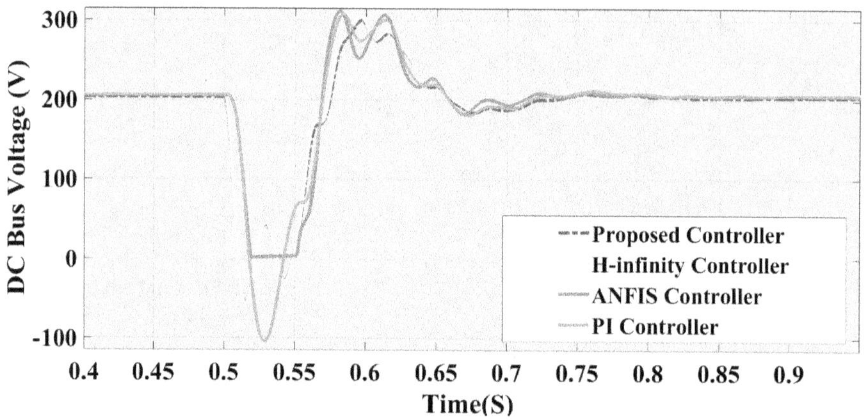

Fig.5.12: Bus voltage during fault

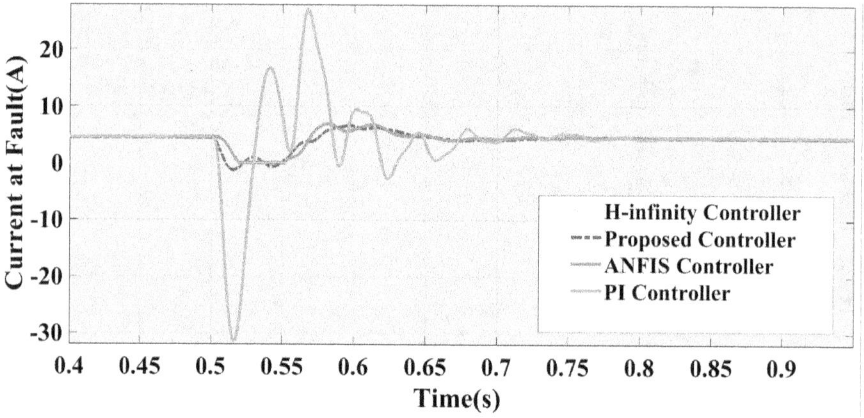

Fig.5.13: Bus current during fault

All of the controllers can stabilise the voltage, as presented in Fig.5.12. However, compared to the suggested controller, the typical PI control has a higher transient response and requires more time to converge. The ANFIS-based approach and the H-infinity controller respond more quickly than the PI controller. However, the suggested approach takes precedence over the other controller when there is a fault. The simulation finding shows that the suggested controller aids the microgrid in fast recovering from these perturbations.

Table 5.2 compares the rise time, overshoot, settling time, and steady-state inaccuracy for various techniques. The recommended controller outperforms the PI, ANFIS, and H-infinity mode controllers concerning overshooting, settling time, rising time, and steady-state errors.

TABLE – 5.2

Various Approaches Time Response for DC microgrid control

Control Design Methods	Rise time(s)	Settling time(s)	%overshoot	% Steady state error
PI	0.052	1.6091	0.326	0.9231
H-infinity	0.047	0.14	0.245	0.21
ANFIS(without observer)	0.044	0.25	0.46	0.21
Proposed design	0.042	0.12	0.22	0.11

5.3 Conclusions:

To achieve efficient voltage management at the DC bus, ANFIS-based controllers were developed in this research study for a DC microgrid powered by renewable energy systems. The major advantage of applying an ANFIS controller is increasing the system's overall dynamic response and regulating the output voltage to ensure dependable system functioning. The control strategy described here is an intelligent control strategy that offers adaptive robust control for the nonlinearities added by the CPLs to address all the challenges effectively and make the system robust against noise and disturbances.

The purpose of this research was to get the bus voltage which tracks to the intended voltage while stabilising a DC microgrid with unknown time-varying loads. To achieve this, the system states vector was enhanced with the uncertain load power, and its spontaneous value was calculated using HCKF. To determine the switch's duty ratio, the predicted power is then feedforwarded to a controller with an ANFIS architecture.

Since CPLs coupled to the DC microgrid create all the nonlinearities and rapidly change the behaviour of the DC microgrid, nonlinear controllers have been considered. To confirm the dynamic stability of the system under the bounds of power situations, Lyapunov stability criteria were also applied. The suggested controller's initial transient reaction is compared to the H-infinity-based, PI-based, and ANFIS-based (without observer) controllers.

Additionally, the output voltage and current at the bus are compared when the load is changed and when an unintended fault occurs. Settling time, rise time, percentage overshoot, and percentage steady state error were also compared for the abovementioned methods.

In comparison to other controllers, the performance of the suggested controller is quite effective, with a settling time (0.042 s), percent overshoot (0.22 percent), and steady-state error (0.11 s). The result is compiled in Table 1.

The experiment is conducted using MATLAB/Simulink tools. The results explain that the suggested control strategy outperforms those put forth by the H-infinity-based controller, the PI controller, and the ANFIS based (without observer) controller. The experiment is performed using MATLAB/Simulink tools, and the outcomes demonstrate that the proposed control strategy outperforms the H-infinity-based, PI-based, and ANFIS-based (without observer) controllers.

CHAPTER 6

HIL SIMULATION:

6. Experimental assessment and demonstration:

This chapter includes experimental case studies that make use of WAVECT to assess the efficacy and reliability of suggested adaptive robust control strategy techniques for DC microgrids under initial transients and fault conditions. Experiments were conducted to validate the findings using the hardware-in-the-loop(HiL) technique. HiL simulation is a well-known prototyping method used to save design costs for power systems by conducting thorough tests early in the development process.

The real-time HiL technique has advantages over conventional offline simulations, including the ability to model delays and faults. The performance of the suggested approach is analysed using the WAVECT control platform. WAVECT assists with programming and real-time data analysis challenges.

WAVECT controller is a Real-Time control prototype system that is portable and flexible. It is made to deal with the complex issues of today. It creates a comprehensive control solution combined with a simple and effective model-based programming environment, pushes button device binary and configuration creation, and specialised instrumentation software. The user equipment is a WCU300. It is located in the research lab, as shown in Fig. 8.1. HiL simulation is a real-time simulation of a physical system that runs at the same rate as the actual physical system.

MODEL:

The MATLAB/Simulink platform is utilised to simulate the Control algorithm. WAVECT supports the HDL coder (comparable to the Simulink blocks) and/or Xilinx System Generator for the modeling of control functions. The WAVECT library will allow MATLAB algorithms to interface with hardware and software. The Simulink library browser offers access to the WAVECT WCU 300 Toolbox as a distinct library.

The interfaces and design are configured using the WAVECT development block. Soft ports are used to provide any external inputs to configure in real-time, while output ports, sometimes known as probes, are used to provide any outputs or internal signals to examine in real-time. On a Simulink model, conversation fields can be used to drag and drop

WAVECT blocks to configure signal parameters. One Subsystem must contain all the functions that need to be coded into the WCU300 Controller (hereafter called Control subsystem).

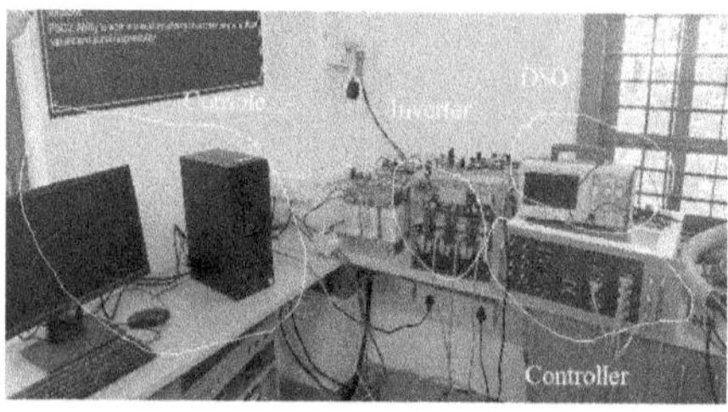

Fig. 6.1: WCU200 Real-time Hardware in the loop　　**Fig. 6.2: WAVECT Controller (HIL)**

Technical specification for WCU200:

- Memory
- 512 MB DDR3
- 256 Mb Quad-SPI Flash
- 4 GB SD card

The control model is transformed into an executable program file when it is programmed into the controller, allowing it to control the physical plant using real-time control signals. Fig. 8.2 depicts the HiL layout built on the Real Time Simulation(RTS).

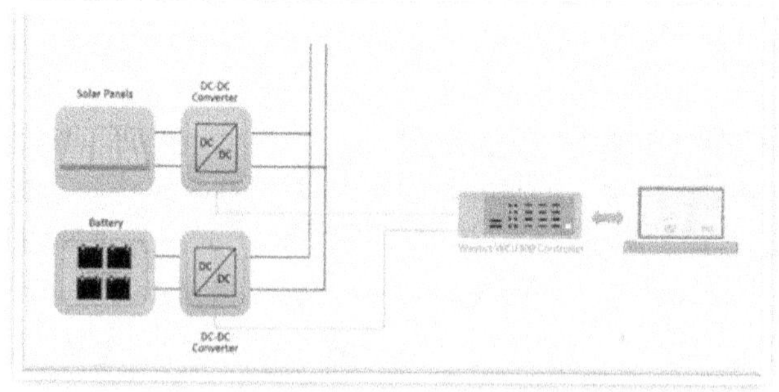

Fig.6.3: WAVECT Microgrid Model for HIL Simulation

It consists of

- Solar PV Panel (330 W each module) - 2 KW

91

- Lead Acid battery   (1 KW, 96V)
- WAVECT WCU300 / WCU300 HD (FPGA Based user programable controller for real-time monitoring and analysis.)
- A PC serving is the programming host (command station) based on MATLAB/Simulink algorithm which generates the WAVECT
- A router serving as a hub for all the setup hardware.

(a) Software

- MATLAB/Simulink - 2017a or higher

- WAVECT WCU300 Toolbox R2021.04 Library

- WAVECT Suite – 2021.04 version

- Vivado - 2018.3 or a compatible version with the installed MATLAB version

- Windows OS - Windows 10, 64-bit

Procedure for Experiments :

The process is divided into four sections.

(i)   Model: Control function modeling is done using MATLAB/Simulink HDL Coder.

(ii)   Simulation: Using the plant model, the control function is simulated to ensure the model performs as predicted.

(iii)   Binary function generation: Control Function binary file generated.

(iv)   Testing: WAVECT Suite is used to load generated binary on the WCU300 Controller and evaluate the model's functionality.

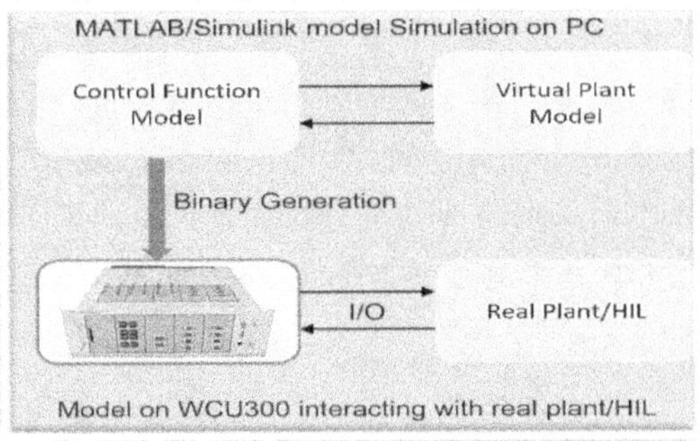

Fig. 6.4: Block diagram for control algorithm testing

A test bench with a solar panel and a battery are built to make sure the efficacy of the design for the suggested control design approach. The solar panel is linked to the DC-bus using DC-DC boost converters. The parameters of the test setup are shown in Tables 8.1 and 8.2. The bus voltage is given as 230 V. Fig. 8.3 displays the control algorithm testing block diagram.

The test setup is seen in a photograph in Fig. 6.2.

TABLE 6.1

Solar photovoltaic module model

Voc	46.10 V
Isc	9.05A
Vmax	37.92V
Imax	8.64 A
Normal operating cell temperature	45°C
Pmax	330W (each module)

TABLE 6.2

Battery Parameters

Nominal voltage	96 V
Rated Capacity	60 AH

The suggested control algorithm is carried out using the real-time simulator WAVECT. The procedures are created in MATLAB/Simulink, afterward compiled and transferred to the WCU300 controller. This facilitates quick and straightforward implementation of the test. The steady-state output voltage is checked using the Multimeter, as shown in Fig. 6.5, to validate the controller output.

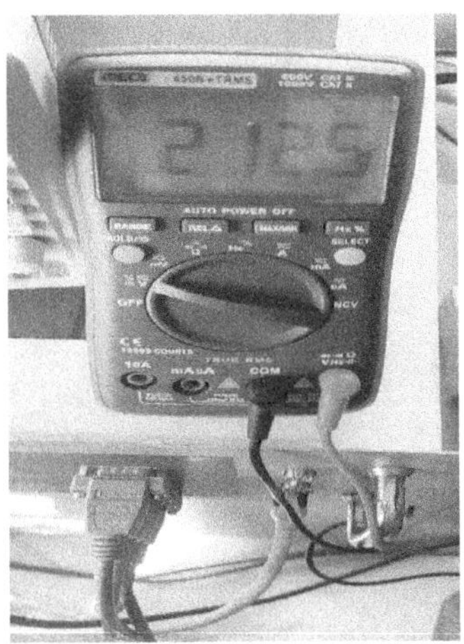

Fig.6.5: Steady state Voltage output

The test is done in two conditions :

Case 1: Initial Transient condition:

In case 1, the experimental validation is carried out during the initial transient using the suggested adaptive robust control approach. The PV system and ESS is operating using the proposed controller. The response during transients is depicted on the graph in Fig. 6.6. The system quickly returns to its initial value after 0.35-second time. The percentage overshoot calculated is 0.08.

A DC microgrid connected to renewable energy sources must be designed with high-power transients as a top priority to assure stability. The system enters an unstable area during transients due to the CPL's increased nonlinearity. The controller should be designed to preserve system stability while ensuring reliable performance during transients. The system states are altered by transients because the CPLs' currents and voltages diverge from their equilibrium points.

The DC microgrid parameters (given in Tables 6.1 and 6.2) are selected to guarantee the system's stability and transient response.

Fig. 6.6 demonstrates that by utilising the suggested controller design, the system can quickly stabilise following early transients and accurately monitor the reference voltage at 200 V.

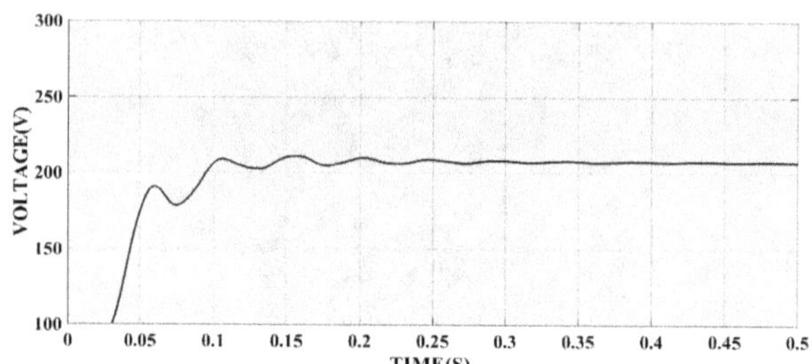

Fig. 6.6: Initial Transient Response

Case 2: Impact of fault

In case 2, the proposed control approach is put to the test in a fault condition. The system is functioning at its nominal specifications, as shown by Tables -6.1 and 6.2. An external fault was introduced after 0.5 seconds after the start. If the fault is not addressed, the system will be forced into an unstable state.

Fig. 6.7 demonstrates that the controller can adjust the current with less time and a minimum overshoot when the disturbance occurs. The graph plotted in Fig. 6.7 demonstrates how the system is reliable and swiftly returns to normal operation following a fault disruption. It shows that using the suggested adaptive robust control approach, the microgrid voltages can quickly return to their nominal value within 0.25 seconds. The steady-state error calculated after fault clearance is 0.06%. The system response shows that the suggested controller manage fast recovery under fault condition.

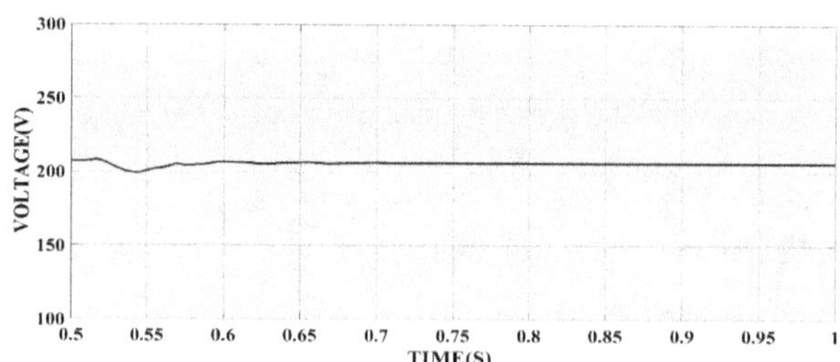

Fig. 6.7: Fault condition Response

The Fig. 6.6 and 6.7 shows that the voltage regulations are reasonable during initial transient and fault conditions.

95

CHAPTER 7

CONCLUSIONS AND FUTURE WORK:

7.1 Conclusions

When traditional energy sources like coal and gas are burnt to power plants, they emit carbon dioxide and other greenhouse gases. On the other hand, renewable energy sources do not emit carbon dioxide when used to generate electricity. By reducing the system's carbon emissions, the globe can achieve the larger objective of net zero emissions. The power distribution network can incorporate cleaner, more effective renewable energy sources through the use of DC microgrids. Therefore, DC Microgrids are a feasible alternative for power systems for offering both technological and environmental benefits.

DC microgrids have numerous advantages over AC microgrids, including fewer power conversion steps and the absence of the requirement for reactive power, phase, or frequency control. DC microgrids have become more popular due to the increased connection of renewable resources like solar PV and chemical batteries, which are essentially DC-type sources. Another significant factor in the growing demand for DC microgrids is due to the increased consumption of DC loads in domestic, commercial, and industrial applications.

Therefore, DC microgrid development has been a significant research area in recent decades to obtain an improved approach to incorporating renewables and ensuring the more consistent function of modern electronic-based devices in power systems. However, integrating distributed energy sources(DER) can challenge the stable operation of DC microgrids as power electronics-based DER have a considerable negative impact on the power system.

These power electronics-based devices act as constant power loads to provide constant power for the device and exhibit negative impedance characteristics. The study demonstrates that DC microgrids can encounter instability because of constant power loads (CPL), resulting in brownouts or blackouts. Therefore, this research aims to provide a suitable controller design to minimise the challenge posed by the CPLs and make the development of the DC microgrid simple and efficient. This dissertation proposes an intelligent adaptive robust control (ARC) method to design controller with high-performance to minimise the negative effect of nonlinear CPLs connected to the DC microgrid.

7.2 Summary:

This chapter compiles a concise summary highlighting the primary development objectives met by this thesis. It discusses several observations and any possible restrictions and deliberate presumptions that may apply to this work. This study proposes a robust adaptive control (RAC) method to address the problem of unpredictable nonlinearities brought on by CPLs.

The proposed method boosts performance by using the combination of deterministic robust control (DRC) and adaptive control benefits. The method can deal with the difficulties of uncertain parameters and guarantees overall transient performance during large signal disturbances.

To begin with, the issue of robust stability analysis for a general DC microgrid with unpredictable CPLs is considered in chapter 3. Since there exist uncertainties, it could be difficult to predict the exact equilibrium the DC microgrid would attain. Therefore, the equilibria that fall into an operationally valuable set are explored initially for a DC microgrid. The formulation relies on selecting a Lyapunov function suitable for the system to attain asymptotic stability. To make the analysis easier, computationally efficient convex problems are developed. The focus is to develop a reliable controller to enhance the stability margin and transient performance of a general DC microgrid while also providing stability analysis.

The challenge is framed as a robust control design with constraints, and an algorithm is developed that ensures the cost functions will constantly decrease. A comprehensive case study is presented using MATLAB/Simulink to show the efficacy of the suggested work. The next step was to examine the adaptive control of nonlinear systems.

A previously established cost function (used in the robust design) for the adaptive model is assessed throughout the implementation of the adaptive controller at each predetermined time interval in chapter 4. The inaccuracy in the mathematical modelling of the systems typically degrades the controller's performance, especially for nonlinear and complicated control issues, which are usually required by classical control theory to construct the controller. Since control design for adaptive control does not necessitate in-depth knowledge of the system, adaptive control has been chosen as the next step for investigation and analysis. The control parameters will adapt on their own to produce excellent performance.

The introduction of neural controllers based on multi-layered neural networks and fuzzy logic controllers (FLC) has sparked fresh ideas for the potential implementation of better and more effective and intelligent control. Intelligent control techniques are now identified as a feasible alternative to traditional model-based controls. Hence, this part addresses the challenge of controlling a nonlinear DC microgrid using an intelligent control method called Adaptive Neuro Fuzzy based Inferencing System (ANFIS).

ANFIS is a method to simulate complex nonlinear mappings utilising fuzzy inference and neural network learning techniques. It is a highly effective method for solving many non-linear control issues. ANFIS improves system performance by enabling more efficient handling of issues like uncertainties or unidentified variations in plant characteristics and structures. Compared to fuzzy-based control design and more conventional approaches like PI-based control design, ANFIS-based control design offers quick settling time while minimising the percentage overshoot. Comparing the ANFIS control design to the PI control design (74%) and fuzzy-logic control design (16%), the control design norm for ANFIS is the least (9.8%).

A prevalent shortcoming in earlier methods is that they assume that the CPL is ideal or that a predetermined limit constrains the power load's unpredictability. To overcome these obstacles, it is necessary to instantly assess the power of the dynamic CPLs. The CPL's powers can be obtained using sensors, which reduces the effects of ripple filtering and increases the output impedance.

Additionally, addition of extra sensors makes the system more complex and raises the expense of purchasing, installing, and maintaining it. So, instead of installing sensors, the current and power of CPLs should be estimated using estimation methods. Therefore, to determine the suitable value of the dynamic power in DC microgrids, the CKF estimator is utilised. Then, DC microgrids with CPLs are stabilised by combining the suggested CKF estimator with the ANFIS-based controller.

The CKF's fast estimation convergence, combined with the stability criterion, results in an output feedback controller with enhanced transient performance. In this work, ANFIS refers to the application of Sugeno-type rules in conjunction with a hybrid learning scheme. The system is tested with 49 rules (7x7) and complete partitioning of the input data space. Following that, to compare ANFIS, a conventional PI controller and a fuzzy-based robust controller design are simulated in MATLAB using the same data.

The MATLAB simulations validate the proposed approach's effectiveness and performance improvements over state-of-the-art controllers. Finally, in chapter 5, a robust adaptive control framework has been developed for nonlinear uncertain systems with modeled uncertainty in the last chapter by combining robust and adaptive control approaches.

In particular, the focus is on an adaptive robust control problem that ensures the DC microgrid's asymptotic stability. The design addresses the challenge of uncertain parameters such as unpredictable fault conditions, sudden and large load variations, and transients. The suggested framework is based on the Lyapunov theorem and captures the remaining prediction errors present in traditional adaptive control techniques.

The approach simplifies the challenge of determining an ARC Lyapunov function for a system's adaptive robust control. A systematic state estimation algorithm for nonlinear systems is proposed by combining the CKF algorithm and H-infinity filtering in the design, and Intelligent control techniques built on the ANFIS are used for the control design.

The stability of the estimation error provided by state observers, typically used to asymptotically estimate the state variables of dynamic systems, has been considered. Also, since no precise microgrid parameters are needed to analyse and design the controller, the suggested method is more useful in actual microgrids.

MATLAB/SIMULINK is used for simulation to establish the effectiveness of the suggested control approach. The suggested controller outperforms conventional PI controllers, H-infinity controllers, and ANFIS-based controllers without state observers during transient response and under various load variations and fault conditions.

Following are some significant contributions made by this dissertation:

- The study suggests improved adaptive-robust control strategies for regulating DC microgrid voltage under various operating conditions.
- Through extensive simulation case studies, the study carefully examines the effects of the suggested adaptive robust control strategies on microgrid stability and dynamic performances. The findings support the effectiveness of the controls, speed up the development of the microgrid, and motivate further methodological advancements.

This thesis offers a comprehensive control design methodology to address the nonlinearity problems caused by CPLs connected to DC microgrids. The suggested methods have the potential to improve the aforementioned systems' effectiveness and performance while guaranteeing that the systems will remain stable across a variety of operations.

As a result, the research's primary goal, as described in this thesis, is satisfied by offering more reliable DC microgrids based on renewable energy that can address the challenges of connecting CPLs.

7.3 Future work:

In order to advance the study, it may be worthwhile to look into the following areas:

- The complex configurations are not considered because this study focuses on a single-bus DC microgrid. There is still work to be done on the multi-time scale modelling approach to other complex topologies.

- This study validated the efficacy of the suggested adaptive robust control for DC microgrid stability. It can be expanded to take into account the effects of a microgrid controller with communication capabilities.

- Other unmodeled dynamics, such as the dynamics of converters, DC sources, or different varieties of nonlinear loads, may be considered in the additional studies.

- The suggested control methodologies can be further examined using the potential reconfigurations between multiple microgrids. The cumulative effects of microgrids on higher-level grid dynamics can be studied for various stages of islanding events or resynchronizations.

- Storage devices operating on various timescales are the foundation for the DC Microgrid's stability. However, the system will be more dependable if additional storage devices, like fuel cells, are added. This will also make it possible to consider economic factors and pursue sustainability goals in addition to dependability and power quality.

- The DC Microgrid's connectivity to an AC grid is also something that will be taken into account. Consequently, the system behaviour and any support services that DC might offer to AC could be better realised. Using a DC grid to bring renewable energy to customers in AC would be very interesting.

9 7 9 8 8 6 8 9 7 5 3 0 1